The Rescue of the *Gale Runner*

NEW PERSPECTIVES ON MARITIME HISTORY AND NAUTICAL ARCHAEOLOGY

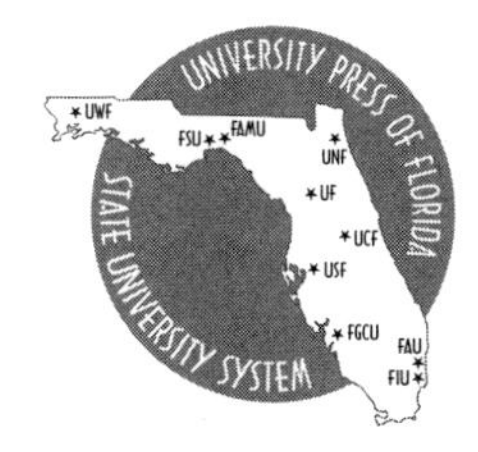

FLORIDA A&M UNIVERSITY, TALLAHASSEE
FLORIDA ATLANTIC UNIVERSITY, BOCA RATON
FLORIDA GULF COAST UNIVERSITY, FT. MYERS
FLORIDA INTERNATIONAL UNIVERSITY, MIAMI
FLORIDA STATE UNIVERSITY, TALLAHASSEE
UNIVERSITY OF CENTRAL FLORIDA, ORLANDO
UNIVERSITY OF FLORIDA, GAINESVILLE
UNIVERSITY OF NORTH FLORIDA, JACKSONVILLE
UNIVERSITY OF SOUTH FLORIDA, TAMPA
UNIVERSITY OF WEST FLORIDA, PENSACOLA

New Perspectives on Maritime History and Nautical Archaeology
James C. Bradford and Gene A. Smith, Series Editors

This series is devoted to providing lively and important books that cover the spectrum of maritime history and nautical archaeology broadly defined. It includes works that focus on the role of canals, rivers, lakes, and oceans in history; on the economic, military, and political use of those waters; and on the people, communities, and industries that support maritime endeavors. Limited neither by geography nor time, the volumes in this series contribute to the overall understanding of maritime history and can be read with profit by both general readers and specialists.

Maritime Heritage of the Cayman Islands, by Roger C. Smith (1999), first paperback edition, 2000
The German Navies after World War II: Dissolution, Transition, and New Beginnings, by Douglas C. Peifer (2002)
The Rescue of the Gale Runner: *Death, Heroism, and the U.S. Coast Guard*, by Dennis L. Noble (2002)

. . . The Rescue of the

Death, Heroism, and the U.S. Coast Guard

University Press of Florida
Gainesville . Tallahassee . Tampa . Boca Raton
Pensacola . Orlando . Miami . Jacksonville . Ft. Myers

Dennis L. Noble

Printed in the United States of America on recycled, acid-free paper

07 06 05 04 03 02 6 5 4 3 2 1

Library of Congress Cataloging-in-Publication Data
Noble, Dennis L.
The rescue of the Gale Runner: death, heroism,
and the U.S. Coast Guard / Dennis L. Noble
p. cm. — (New perspectives on maritime history and nautical archaeology)
Includes bibliographical references and index.
ISBN 0-8130-2555-9 (cloth: alk. paper)
1. Gale Runner (Sailboat). 2. Quillayute River (Wash.: Coast Guard Station).
3. United States. Coast Guard—Search and rescue operations—Washington (State).
4. Shipwrecks—Washington (State). I. Title. II. Series.
VK1255.G35 N63 2002
910'.9164'32—dc21 2002018076

The University Press of Florida is the scholarly publishing agency for the State University System of Florida, comprising Florida A&M University, Florida Atlantic University, Florida Gulf Coast University, Florida International University, Florida State University, University of Central Florida, University of Florida, University of North Florida, University of South Florida, and University of West Florida.

Cover image: "February 12, 1997," by W. Brent Cookingham

University Press of Florida
15 Northwest 15th Street
Gainesville, FL 32611–2079
http://www.upf.com

This book is for George A. LaForge, Jonathan A. Placido,
Benjamin F. Wingo, and the 12 February 1997 crew of
the Quillayute River U.S. Coast Guard Station, survivors all,
and
to all those in the U.S. Coast Guard who have pushed out
into high seas in small boats so that others might live.

"death knows . . . where people live who do dangerous work for not much money."

Thomas Harris

"The small boat community of the U.S. Coast Guard is a world of so much passion and intensity to do the right thing."

CWO2 F. Scott Clendenin
Commanding Officer, Yaquina Bay U.S. Coast Guard Station, Oregon

"We ask so much of our young people in uniform and they so freely give it."

BMCM George A. LaForge, U.S. Coast Guard (retired)
Former officer-in-charge, U.S. Coast Guard Station, Quillayute River, Washington

Contents

Part Three. Aftermath

Figures, Maps, and Table

Figures

Maps

Table

Foreword

Water is unquestionably the most important natural feature on earth. By volume the world's oceans compose 99 percent of the planet's living space; in fact the surface of the Pacific Ocean alone is larger than the total of all the land bodies. Water is as vital to life as air. To test whether the moon or other planets can sustain life, NASA looks for signs of water. So the story of human development is inextricably linked to the oceans, seas, lakes, and rivers that dominate the earth's surface.

While water is necessary for creating and sustaining life, it can also take lives, as it did in the Pacific Northwest off the coast of Washington on the night of 11–12 February 1997. During that evening the weather quickly turned severe, catching the sailing vessel *Gale Runner* unprepared. The crew of the sailboat transmitted a "Mayday" distress signal while trying to put into a safe harbor. The Coast Guard's Quillayute River small-boat station received the signal and immediately put their 44-foot motor lifeboat into action to rescue those in peril. In the dramatic search-and-rescue operation that followed, nothing could have prepared the crew members for what they encountered in the darkness and the rough-and-tumble seas. Coast Guard lifeboats and helicopter were pounded incessantly as the crews did all within their power to save the troubled sailing vessel. Both the occupants of the sailboat and their would-be rescuers suffered severely—three members of a four-man lifeboat died when their vessel capsized and was tossed upon the rocks—but the two people aboard the sailboat survived without serious injury. The death of three Coast Guardsmen that evening was the worst loss of life experienced by a Coast Guard small-boat station since 1961, and Dennis Noble's *Rescue of the Gale Runner* is their dramatic and gut-wrenching story.

Dennis Noble is well qualified to tell this story. A retired Coast Guard marine science technician, he earned a Ph.D. in American history at Purdue University and is the author of several books concerning the Life-Saving Service and Coast Guard, including *That Others Might Live: The U.S. Life-Saving Service, 1878–1915* (1994) and *Lifeboat Sailors: Disasters, Rescues, and the Perilous Future of the Coast Guard Small Boat Stations* (2000). More important, Noble was present at the Quillayute River Station on the night of the tragedy and saw firsthand the actions and emotions of those involved. Noble's presence on the scene that night, his previous research, and his service in the Coast Guard provide him with a unique perspective. By blending historical scholarship, investigative reporting, governmental watch-dogging, and first-person oral history, Noble places the accident in context, describes the event and its consequences, and suggests its implications for the Coast Guard's present and future.

Noble makes several important contributions in this book. He describes the dangerous and thankless work of the men and women of the U.S. Coast Guard's small-boat stations; he distinguishes the considerable differences between the officers and enlisted personnel who conduct search-and-rescue missions and those far-removed administrators who make policies that affect both the Coast Guard units and the boating public; and he memorializes all of those who played a role in this powerfully dramatic affair. His book assures that the Coast Guardsmen who participated in the rescue, including those who lost their lives saving others, will not be forgotten.

James C. Bradford and Gene A. Smith, series editors

Preface

The purpose of this book is to shed light on a group of people who, since 1878, have rammed small rescue boats into high seas so that others might live. All too often the deeds of these men and women have gone unheralded. "The sea," wrote Joseph Conrad, "has no generosity." Operating small rescue boats in gales and high waves is an inherently dangerous profession. One wave that does not act the way others do, one minute of inattention, one look in the wrong direction, or one wrong decision can spell death. The writer Paul Theroux observed, "The sea in winter is unforgiving—much worse than any wilderness of snow-laden trees, more merciless that a mountainside, harsher than any desert. If you go down in cold water and lose your boat, you are doomed as soon as your core temperature drops—and it drops in minutes in a winter sea."

On 11–12 February 1997, the men and women of the isolated U.S. Coast Guard Station Quillayute River, located in LaPush, Washington, found how uncaring the sea can be. How these men and women, and others from the aviation branch of the U.S. Coast Guard, met the sea is a story of heroism that novels and Hollywood fail to match.

I have tried to let the people who experienced the events of 11–12 February and the aftermath tell the story as much as possible. Readers may find some contradictions as to times, and even different observations of the same event, but each is what an individual felt or actually observed. Military historians have long written on the "fog of war." The events taking place during a major search-and-rescue case in very bad weather—and most seem to occur only in bad weather in the dark of night—could be a textbook model for the fog of war. This phenomenon explains why there can be confusing observations of the same event. I hope the many voices in this book will show just how confusing things can seem.

Few Americans understand anything about the people who risk their lives to perform maritime rescues. Because of this lack of information, I have added explanatory passages throughout the text about the people and their equipment.

The people at the Quillayute River station also met another danger: the atmosphere after their combat with the sea. Questions, innuendo, and gossip plagued the survivors, all part and parcel of the story.

Those who read this account may point out that some of those involved might have made mistakes. To those in a warm room and in an easy chair, it is one thing to point a finger; it is quite another to make a decision when the wind is howling, the rain is coming at you horizontally, and someone is yelling "Mayday!" over a radio. The people in this story are human and have human foibles.

In a previous work on the small-boat rescue stations of the U.S. Coast Guard, I wrote at length about how the decision-makers at Coast Guard headquarters seem to care little about the people who put their lives on the line at the small-boat stations. I also noted how many in the enlisted force were concerned that the U.S. Coast Guard's senior leadership would punish them for speaking out against certain policies. The incident at Quillayute River and the aftermath has confirmed my views. Now, as then, I have protected the names of those who are most at risk.

One crew member pointed out that anyone closely involved with the Quillayute River incident "has an albatross around their necks," but, the crewman pointed out, there are some who have not acknowledged any blame in the incident. I will leave to the reader the decision as to who failed to shoulder any responsibility for what happened on 12 February 1997.

Some have pointed out that because I was at the station during 11–12 February, I cannot be objective. So that readers may understand my bias, I spent over 20 years in the U.S. Coast Guard, serving as an enlisted man. I grew to think of the people at the Quillayute River station as my friends and myself as a part of their "family." (I still enjoy visiting with the current crew at Quillayute River.) Furthermore, my writings on the history of the U.S. Coast Guard show that my heart is with the enlisted force that has been largely overlooked. I have tried to obtain opposing views and place them in the narrative so that readers may make up their own minds on the subject. My own opinions are clearly stated. In this work, the reader will see some officers performing heroic deeds, but, not to diminish their efforts in any way, the focus of this account is on the enlisted force.

Many will see portions of this book as a diatribe against the officer corps of the U.S. Coast Guard. A few officers of the service are outstanding, and a few are less than satisfactory. The great majority, like everyone else in America, are average and try to do their work the best they can. There are those in the officer corps who wish to make good, constructive changes in the small-boat community but are frustrated by their lack of success. Other officers work against almost overwhelming odds to make an organization run on what amounts to a pittance. The American taxpayer should be thankful there are such dedicated individuals in the U.S. Coast Guard.

Among other things, this book is about responsibility for actions. There are officers in the U.S. Coast Guard who know that some policies concerning the stations are wrong, but they do not speak out because of what has been dubbed "career fear." Others in leadership roles care only for the privileges of the officer corps and do nothing about problems that have festered for decades. There are also those who do not care to recognize problems but would rather accept what they are told. Eventually, someone is responsible. In the U.S. Coast Guard, and other military organizations, the members of the officer corps are considered the ultimate leaders in the service, so they must also bear the burden of responsibility for what is both good and bad within the service. This does not mean all officers are uncaring or bad; it simply means someone must be held responsible.

Readers may wonder why I did not dedicate this work to the three U.S. Coast Guardsmen who died. The three men received a dedication in my work on the small-boat stations, and I felt it was now time to recognize the rest of the crew for their efforts that night and in the long days and months that followed.

As noted, a great deal of the material for this book comes from the people who took part in the events of the night of 11–12 February 1997 or from my personal observations. All quotes from people, whether by interview, letter, telephone conversation, or e-mail, are simply put in quotation marks with no footnotes. Likewise, material from the official investigation into the incident is not footnoted. All material from other sources is credited in the notes and bibliography. The words of the participants are as they spoke them. Some may be grammatically incorrect, but then, most people do not speak in perfectly grammatical English. I have edited out repeated material, false starts, and other hitches that occur in everyday speech. The abbreviations for ranks and rates are those used by the U.S. Coast Guard in 1997.

Chief Warrant Officer F. Scott Clendenin, commanding officer of the Yaquina Bay, Oregon, station, said, "The small-boat community of the U.S. Coast Guard is a world of so much passion and intensity to do the right thing." If the reader comes away with an understanding of the events of 11–12 February 1997 and of the people who have this "passion and intensity to do the right thing," the book will have served its purpose.

I finished this book at the U.S. Coast Guard station Quillayute River, Washington, the station that is at the heart of this story.

This book is based largely on the observations of many people. Instead of focusing on any one individual I interviewed, I would like to thank all for their contributions. To dredge up memories that were not always the most pleasant required a great deal of fortitude, and it seems very little to say thank you.

Once again, Dr. Robert M. Browning, Jr., Historian of the U.S. Coast Guard, and his staff of Scott T. Price and Christopher Havern, went out of their way to assist me in my work. The U.S. Coast Guard is fortunate to have such dedicated professionals.

At U.S. Coast Guard Headquarters, Captain David Kunkel provided me with material on the HH-65A helicopter and information on the *Marine Electric* case. Captain Kunkel has been very supportive, even prior to his headquarters tour while serving as commander, Group/Air Station Astoria, Oregon. Master Chief Aviation Survival Technician Keith R. Jensen, Helicopter Rescue Swimmer Program manager, provided me with needed material on the training and outfitting of rescue swimmers.

Master Chief Petty Officer of the Coast Guard Vincent W. Patton III helped a great deal by responding to questions when I could not get a response from headquarters. Whenever he could, he cut through red tape to help me.

U.S. Coast Guard headquarters passed on many of my questions, if not all, to the Thirteenth Coast Guard District in Seattle for their responses. Captain William W. Peterson and his staff of Lieutenant Commander Andrew Connor and Master Chief Boatswain's Mate Curtis W. Mauck were given this duty. Both officers, while disagreeing with many of my points, responded quickly to my questions. I am certain that it was not pleasant having to take time out of their busy schedules to answer some pointed questions that they disagreed with, but they met the challenge with a great deal of professionalism. I have heard many officers-in-charge in the small-boat community comment on how Captain Peterson has tried to help them, and I can see why they have passed this information on to me. Chief Public Affairs Specialist John Moss, also at the Thirteenth District, provided me with photographs.

Captain Gene Davis, U.S. Coast Guard (Retired) at the Coast Guard Museum Northwest in Seattle, provided me needed information and photographs.

Captain Philip C. Volk, U.S. Coast Guard (Retired), has been one of my largest supporters. While serving as commander of the U.S. Coast Guard Group/Air Station Port Angeles, Washington, he helped make my work at the small-boat stations of the service possible. As the man who made many decisions during the long predawn hours of 12 February 1997, Captain Volk's frank and insightful observations helped greatly in this study. He is the type of officer with whom enlisted people hope to serve.

Lieutenant Commander Robert W. Steiner, of U.S. Coast Guard Group Astoria, helped a great deal in obtaining information on the personnel of 6003. Lieutenant Daniel C. Johnson, commanding officer of the Cape Disappointment Station, and Boatswain's Mate 1 Jeff Kihlmire helped greatly in obtaining the materials for the story of the *Miss Renee* case. Lieutenant David C. Billburg, public affairs officer, U.S. Coast Guard Group/Air Station Port Angeles, quickly provided me with much needed material on the helicopters attached to the Port Angeles Air Station and other materials concerning the group/station. Chief Warrant Officer 4 Mark Dobney and Boatswain's Mate 1 Darrin Wallace of the U.S. Coast Guard National Motor Lifeboat School, Ilwaco, Washington, responded quickly and efficiently to questions on motor lifeboats.

Chief Warrant Officer 2 Thomas Doucette, U.S. Coast Guard (Retired), read the manuscript, offering me invaluable comments, both as a surfman and as a crewman at the Quillayute River station early in his career. Over the years his insights and humor have been of great help in my work.

I have been lucky enough to have known Master Chief Boatswain's Mate Daniel Shipman, U.S. Coast Guard (Retired), while he served as officer-in-charge of the Quillayute River and the Tillamook Bay, Oregon, stations. His candid comments were extremely useful to this work. Chief Warrant Officer 2 F. Scott Clendenin, U.S. Coast Guard (Retired), took time out of his very busy schedule to discuss many items with me. Scotty's retirement marked the end of an era for the small-boat community, and I shall miss seeing him at the Yaquina Bay, Oregon, station.

Senior Chief Boatswain's Mate Dave Mayrick, the current officer-in-charge of the Quillayute River Station, and his executive petty officer, Chief Boatswain's Mate Michael Saindon, were always there to help me, despite their many other duties. Mike especially made a point of being at the unit when I arrived, even going so far as to come in from home to speak with me. The crews of the Quillayute River Station from 1997 to the present, despite their isolation and very long work hours, have always gone out of their way to make me feel at home. They have never complained about the extra times they got under way so I could better understand the sea around James Island and The Needles. The Quillayute River crews that I have known over the years represent what is best about the small-boat community.

I would also like to thank all those who, when they knew headquarters had "lost" items, went out of their way to make sure I obtained what I needed. Much of what they provided helped to shed light on many questions that could not have been answered otherwise.

Peggy Norris read the manuscript and once again used her outstanding abilities to spot inconsistencies and, of course, my misspellings. The editor of this work should thank Peggy for her work. Loren A. Noble read the manuscript and offered helpful comments. When one-third of the manuscript had to be retyped because of the author's stupid mistake with his nemesis, the computer, Loren assisted in the retyping, thus speeding up the completion of the chore. Susan Browning produced the outstanding maps for this work. She continues to make my work look good. Tom Beard, helicopter pilot, sailboat sailor *extraordinaire,* maritime historian, author, and friend, read the manuscript and lent his expertise to my work. He also endured my ranting and raving about the manuscript through so many lunches that he should be made an honorary member of the Quillayute River crew. Greg Shield took time out from his own work to read the manuscript and offer very insightful comments. He had no idea he would venture out into such a roiled sea of split infinitives.

William D. "Bill" Wilkinson, director emeritus of The Mariner's Museum, helped me greatly with the history of the lifeboats used by the U.S.

Coast Guard. He is the recognized international expert on the motor lifeboats of the U.S. Coast Guard and has always been a source of encouragement and information over the years.

The University Press of Florida's series "New Perspectives on Maritime History and Nautical Archaeology" is edited by James C. Bradford and Gene A. Smith, and I wish to thank them for their help. Gene Smith deserves special recognition for his insightful and helpful comments. Meredith Morris-Babb steered the manuscript to its completion.

Where this work has merit, it is because of all these people; where it falters is where I disregarded their sound judgment and advice.

Abbreviations/Terms

Ranks/Rating Abbreviations and Terms used in the U.S. Coast Guard along with the Chain of Command

Officers	Pay grade	Title
ADM	O-10	Admiral
VADM	O-9	Vice Admiral
RADM	O-8	Rear Admiral (Upper half)
RADM	O-7	Rear Admiral (Lower half)
CAPT	O-6	Captain
CDR	O-5	Commander
LCDR	O-4	Lieutenant Commander
LT	O-3	Lieutenant
LT (j.g.)	O-2	Lieutenant (junior grade)
ENS	O-1	Ensign
Warrant Officers		
CWO-4	W4	Chief Warrant Officer 4
CWO-3	W3	Chief Warrant Officer 3
CWO-2	W2	Chief Warrant Officer 2
WO-1	W1	Warrant Officer 1

Enlisted Noncommissioned Officers

MCPO	E-9	Master Chief Petty Officer
SCPO	E-8	Senior Chief Petty Officer
CPO	E-7	Chief Petty Officer
PO1	E-6	Petty Officer First Class
PO2	E-5	Petty Officer Second Class
PO3	E-4	Petty Officer Third Class

Nonrated personnel

SN/FN/AN	E-3	Seaman/Fireman/Airman
SA/FA	E-2	Seaman Apprentice/Fireman Apprentice
SR/FR	E-1	Seaman Recruit/Fireman Recruit

U.S. Coast Guard Enlisted Ratings Appearing in this Book (ratings are as used in 1997)

AM	Aviation Structural Mechanic
ASM	Aviation Survivalman
ASTCM	Master Chief Aviation Survival Technician
BM	Boatswain's Mate
BMCM	Master Chief Boatswain's Mate
BMCS	Senior Chief Boatswain's Mate
DC	Damage Controlman
FS	Food Specialist
MK	Machinery Technician
PA	Public Affairs Specialist
QM	Quartermaster
TC	Telecommunications Specialist

Terms Used at Stations

AOR	Area of responsibility
Boot	New person, usually right out of basic training but can mean any new person at a station
Breaks	Breaking waves, or surf
CISM	Critical Incident Stress Management

CO	Commanding Officer. A commissioned, or warrant, officer in charge of a unit
CVSB	Commandant's Vessel Safety Board
Detailer	Officer in U.S. Coast Guard headquarters who makes duty station assignments
Dream sheet	Form that U.S. Coast Guard personnel submit to detailer with their choices of next duty assignment. The official name of the form is the ADC, Assignment Data Card.
EMT	Emergency medical technician
EPO	Engineering petty officer
FLIR	Forward-looking infrared radar
GAR	Green, amber, red
GDO	Group duty officer
GPS	Global Positioning System. A positioning system that gives a quick electronic location. The system receives signals from orbiting satellites.
JOOD	Junior Officer of the Day
LORAN	Long Range Aids to Navigation
NDRSMP	National Distress Response System Modernization Project
Nonrate	Entry-level enlisted personnel in the deck, engineering, or aviation fields who have not obtained a noncommissioned rating
OER	Officer Evaluation Reports
OIC/OINC	Officer-in-charge. Usually a CPO, SCPO, or MCPO.
OOD	Officer of the day
OPS	Operations officer. A commissioned or warrant officer in charge of operations at a unit
OPSPO	Operations petty officer.
Petty officer	An enlisted noncommissioned officer
PIW	Person in the water
Plank owner	A person who is one of the original crew when a ship or station is placed into commission.
PMS	Preventive maintenance schedule

RHIB	A small rigid hull inflatable boat powered by outboard motors
SAR	Search and rescue
SDO	Senior duty officer
SRB	Surf rescue boat
STAN Team	Standardization Team
TCT	Team Coordination Training
TAD	Temporary additional duty
XO	Executive officer. A commissioned or warrant officer who is the second in command at a unit.
XPO	An enlisted petty officer who is second in command of a unit; executive petty officer

Chain of Command at the Quillayute River Station

U.S. Coast Guard Station Quillayute River, LaPush, Washington,
reports to,
U.S. Coast Guard Group Port Angeles, U.S. Coast Guard Air Station, Port Angeles, Washington,
who reports to,
Commander, Thirteenth Coast Guard District, Seattle, Washington,
who reports to,
Commander, U.S. Coast Guard Pacific Area, Alameda, California,
who reports to,
U.S. Coast Guard Headquarters, Washington, D.C.

Introduction

In the dim early morning light of 12 February 1997, members of the Quileute Native American tribe stood on an overlook gazing across the high waves of the North Pacific Ocean. A red U.S. Coast Guard helicopter hovered and darted like a giant seabird around James Island in northwestern Washington State. Many in the tribe knew something had happened during the gale-swept night. Throughout the hours of darkness, word of red flares to seaward had spread throughout the reservation's small village of LaPush. There were rumors of activities in the area of First Beach. The sounds of motor lifeboats could be heard from the U.S. Coast Guard's Quillayute River station, located on the reservation, and much activity at the small station could be observed. Later, the unmistakable high-pitched whine of U.S. Coast Guard helicopters pierced the stormy night. During the previous week, the U.S. Coast Guard station had recovered the body of a tribal member who had drowned. Death is an ever-present part of life for a sea people. Tribal members stood in the strong wind under a leaden sky and driving drizzle, awaiting information on who had died.

This book is an account of the men and women of the U.S. Coast Guard station Quillayute River, Washington, who in the early morning hours of 12 February responded to a call of distress from two people aboard a sailboat. Within a two-hour period, three U.S. Coast Guardsmen were dead, two more had received injuries, one motor lifeboat was damaged beyond repair, a second motor lifeboat was damaged, two people were rescued from a sailboat, and two helicopters came close to crashing.

I happened to be visiting the Quillayute River station the night of the incident. After a period of time to allow for the shock and grieving to recede, I realized the deaths presented a chance to tell the story of the men and women who are expected to go into harm's way to help those in peril on the sea but who rarely receive the accolades they deserve. This story is based on interviews with the people involved in the accident, those who had anything to do with the case, those in the U.S. Coast Guard who know about the station, those in the decision-making structure of the service, and a few personal observations.

An important part of the story is the work of U.S. Coast Guard Air Stations Port Angeles, Washington, and Astoria, Oregon. The pilots of the helicopter and the rescue hoist operator of one of the helicopter crews from Port Angeles won the Distinguished Flying Cross for their work, the highest medal for valor a helicopter crew can earn in peacetime.

Part of this story takes place after those harrowing early morning hours. I have included the results of the investigation into the deaths and the impact it had upon the people in the account, as they are an important part of the whole story.

This is a story of a group of ordinary men and women, who, for a few hours during the height of a howling gale and mountainous seas, rose above the occasion. Those in the higher echelons of the U.S. Coast Guard's command structure will no doubt point out that training prevailed. In a measure, this is true, but this small band of people went above and beyond this simple view. This is their story.

. . . Part One Tuesday, 11 February 1997

1 The Fortress of Solitude

Lights snapped on at 6:00 A.M. on Tuesday, 11 February 1997, in the barrack rooms of the U.S. Coast Guard small-boat rescue station, Quillayute River, Washington.[1] People in the starboard duty section started their normal routines. The section would be relieved at 1:00 P.M., by the port section. The Quillayute River station is located in a region many people in the largely urbanized United States would find difficult to believe still exists in the 21st century.

The Quillayute River station is located at the northern portion of the Olympic Peninsula on the Quileute Reservation (the spelling of the tribal name differs from that of the river and station) near the northwestern tip of Washington State. The unit is considered one of the three most isolated small-boat stations in the U.S. Coast Guard.[2] The region contains an "improbable physiographic assemblage of high-and-mighty mountain peaks, creeping glaciers, shadowy rain forests, brooding lakes, ferocious rivers and storm-battered shoreline."[3] The 5,700 square miles that make up the peninsula are sparsely populated with small towns nestled among large stands of Douglas fir. Clallam County, the state political unit in which the station is located, measures 1,745 square miles yet had a population in 1997 of only 66,000. In the same state, King County, in which Seattle is located, has 2,126 square miles and a population in 1997 of 1,646,200. The population density in Clallam County is 38 people per square mile, in King County 774 per square mile.[4]

Machinery Technician First Class (MK1) Bruce Mumford, the station's engineering petty officer (EPO),[5] said, "We must drive one and a half hours, one way, to go to a movie theater." LaPush, the reservation's village, does not have a theater. Forks, the next nearest settlement, with a population of around 1,000, is about a 20-minute drive from the station and does not have a cinema. The only theater for the crew of the station to relax at is in Port Angeles, the largest settlement on the North Olympic Peninsula, with a population of 19,000. To see a movie a crew member must therefore allot at least three hours of travel. For those with teenagers and families who love to hang out or shop at a mall, the nearest mall is an hour's drive east of Port Angeles, or approximately two and one-half hours away from the station (map 1).

The Quillayute River station is one of the few U.S. Coast Guard units in the United States located on a Native American reservation. The Quileute are Chemakuan speakers, a language group unique to the peninsula, "one of five languages in the world that has no nasal sounds (*m* or *n*)."[6] Prior to the arrival of Europeans, the tribe enjoyed one of the most varied ways of life on the peninsula. Enormous runs of steelhead and salmon filled the rivers, and the Quileutes occasionally also hunted whales and seals. They were noted for their highly decorative artwork, and their warriors had a reputation for fierceness, conducting raids and taking retribution parties along the coast from Vancouver Island in British Columbia to the mouth of the Columbia River.[7] LaPush is accurately described by a tribal publication as an "isolated village largely untouched by the influences of mainstream society."[8]

Over the years, rumors of bad feelings between the station and tribe have been standard. Many former crew members have commented that the "only reason the tribe tolerates the station is that we provide an EMT [emergency medical technician] service and fire-fighting service." It is somewhat unusual to walk into a U.S. Coast Guard station and see such fire-fighting gear as turn-out coats, helmets, oxygen masks, and bottles ready for use as if the unit were a fire station. "There may have been bad relations in the past with the tribe," said Boatswain's Mate First Class (BM1) Michael "Mike" Saindon, in 1999 the executive petty officer

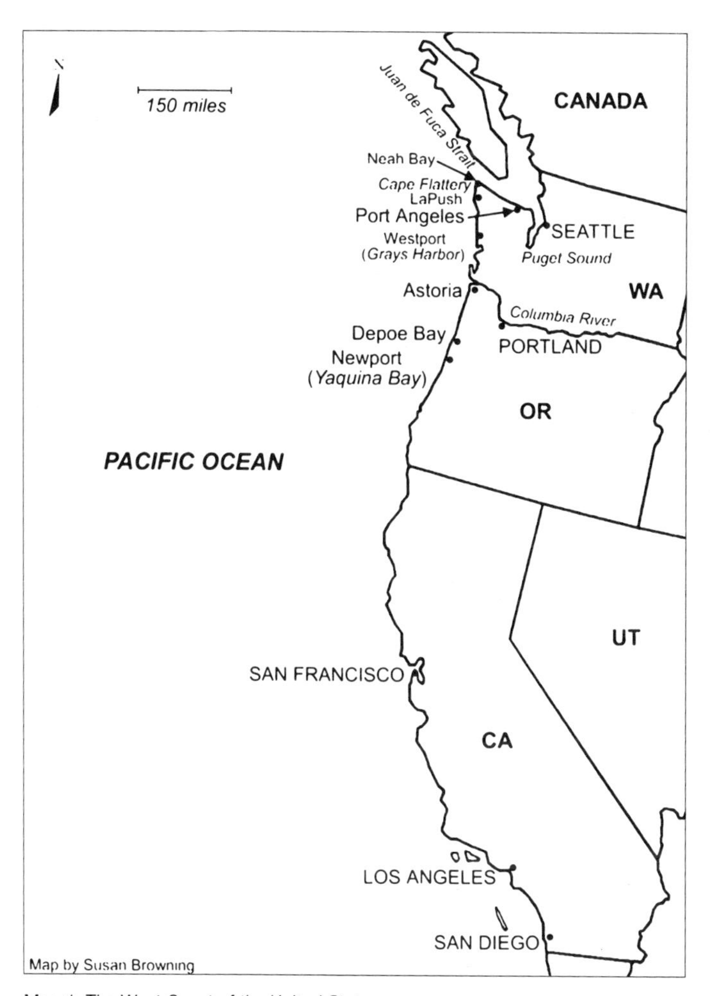

Map 1. The West Coast of the United States

(XPO) of the Quillayute River Station. "We have very good relations with the tribe. In the past, you could usually trace the bad relations to teenagers. Normally, the elders and youngsters like us. There was a case recently of a bad car accident and even though the tribal person was closer to Forks, they elected to bring the person to the station because they are used to us."

Part of the reason for the lack of population in this northwestern location is that at the center of the peninsula Olympic National Park contains about 900,000 acres of land, while the U.S. Forest Service manages another 700,000 acres surrounding the park; thus over a million acres of land has been withdrawn from development. It was not until 1935 that a 325-mile highway, an extension of U.S. 101, finally looped around the peninsula (map 2). This road should have heralded more roads and people. The Olympic Peninsula, however, is different. No road runs completely through Olympic National Park, and any road in the region usually branches off of U.S. 101 and then dead-ends, especially in the northern part of the peninsula where the Quillayute River station is located.

The region contains some of the most varied and marvelous scenery in the United States. Mount Olympus, at 7,965 feet, is the highest on the peninsula—not tall as mountains go in the western United States, but seen from upon the waters of the Strait of Juan de Fuca the peaks seem to rise straight out of the sea. Tourists from Port Angeles, visiting Hurricane Ridge, one of the areas in the high country set aside for visitors, travel from sea level to over 5,200 feet in approximately 17 miles.

The waters of Puget Sound border the eastern side of the peninsula, separating it from the large population centers of Seattle and Tacoma. To approach the northern Olympic Peninsula from the east, you must take a ferry and then, if you want to drive to Quillayute River, cross a floating bridge. From Seattle to Port Angeles, a distance of about 70 miles, takes one and a half hours, if connections are just right, then another hour-and-a-half to two-hour drive. In the summer, with many tourists visiting the park, it is not unusual to wait in line for an hour to get on the ferry. Another route, around Puget Sound's Hood Canal, will take about three hours to Port Angeles. Using the only other route, around the southern boundary of the park and then up the western side, it will take about four hours to reach the Quillayute River station. The northern boundary of the peninsula is the cold waters of the Strait of Juan de Fuca, separating the United States from British Columbia, Canada. The western side is dominated by the Pacific Ocean.

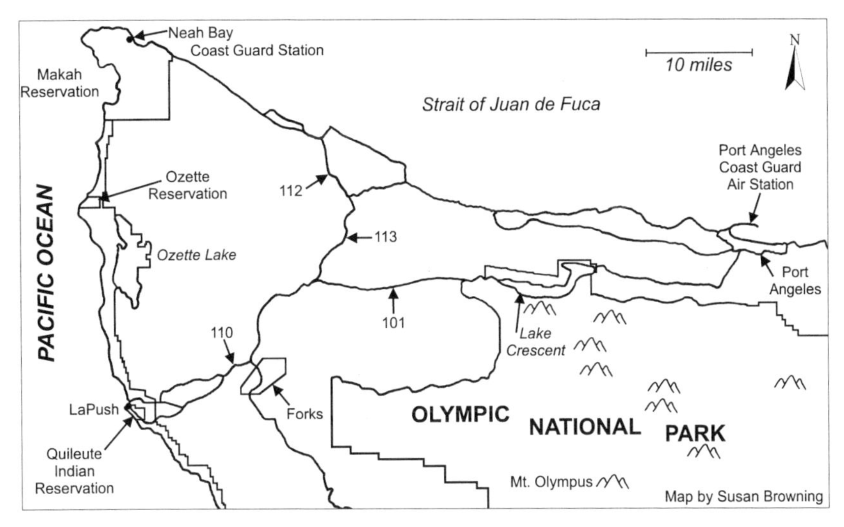

Map 2. The North Olympic Peninsula

In addition to mountains, glaciers, and old-growth forests, the northern coastal area of the peninsula is located within a temperate rain forest, the only one outside of Chile and New Zealand. Lush growths of Sitka spruce, mosses, and ferns make the region appear almost jungle-like. Some of the dark forests can make visitors think of Hansel and Gretel or perhaps of a wood spirit appearing from a hollow moss-covered log. One writer has aptly summed up the peninsula: It "can be a perilous place where if you wander off a forest trail you may never be seen again; yet it is a romantic place where the sight of a delicate ice-age flower poking up through snow, or a 50-pound Chinook salmon hurling rock falls to get upstream to spawn and die, can make you catch your breath."[9] Indeed, over five million visitors came to marvel at this region in 1997, most arriving from May through October.

For those who enjoy wildlife, the region has few equals. As it is located along the Pacific flyway, large numbers of waterfowl and other birds, sometimes hundreds of thousands, may be observed from the front door of the Quillayute River station. Because all inside spaces of U.S. Coast Guard stations are now smoke-free, smokers usually congregate in a designated area near the north side of the breezeway. "I like to watch a group of bald eagles in the trees near the river when taking a cigarette break," said Boatswain's Mate Second Class (BM2) W. Brent Cookingham. The park is the home to the largest herd of Roosevelt elk in North America. Fishing, although not as great as it used to be, is still some of the best in the country. For those who love to hike, fish, and hunt, the region around Quillayute River can make an assignment here an acceptable duty station.

Many of the region's visitors head for the ocean shoreline. While the southern part of the park's ocean strip does contain some sandy beaches readily accessible by car, most of the shoreline is seen only after hiking varying lengths through the temperate rain forest. Second Beach, in the park and within five miles of the Quillayute River station, for example, requires a half-mile hike through a fairly flat temperate rain forest, then a 120-foot slope down to the beach area. Upon reaching the bottom of the grade, the walker must negotiate a log-strewn area, a feature of most beaches in the area. Some trunks have circumferences larger than 5 feet and reach lengths of 50 to 100 feet. The beaches are backed by high precipices. Where the rocky bluffs have a relatively soft rock, the relentless sea has worn away the cliffs. If the remaining material is extremely hard, the sea has formed small islands, or sea stacks, of incredible shapes and sizes, many reaching hundreds of feet into the air. The sea has formed a rocky but incredibly beautiful and wild coastline.

The Pacific Ocean dominates everything in the coastal northern portion of the Olympic Peninsula, especially the weather. Visitors who flock to the park in the summer months comment upon the cool temperatures. It is not unusual for visitors from other locations in the United States to leave their homes in 90-degree summer heat and fly to Port Angeles to find the thermometer 20 degrees lower and the locals saying that it is a warm day. The offshore Japanese current and westerly winds ensure a mild maritime climate. Contrary to a belief held by most people in the United States, in summer it does *not* rain all the time. Wise visitors to the region time their stay for the months from July through October, when there are usually more sunny days than rainy skies.

February at the Quillayute River station is something else. The millions who flock to the park are gone. At the Mora Beach ranger station, fewer than 20 miles from the Quillayute River station, the number of cars passing in August 1996 was 13,787; the count dropped to 4,582 in February 1997.[10] The closer you come to the coast, the more it rains. Summer naturalists in the park have the following formula for the sunshine visitors: There is an additional inch of rain for every mile westward from Port Angeles. Port Angeles's normal rainfall is 25.78 inches per year, and it is approximately 70 road miles from Port Angeles to the Quillayute River station. Rainfall at Quillayute River is 102.48 inches per year. As may be seen, the amount of precipitation at the station is easier to measure in feet than in inches. It is not unusual in the winter months to go for 30 days with low, gray overcast skies. MK1 Mumford's comment is typical of many at the unit: "I get so tired of the constant rain and gray skies. It can really drag you down."[11] The sea water temperature throughout the year ranges from 45 degrees in winter to 57 degrees in summer, so that even in the summer months one is not apt to jump into the water wearing a bikini.

During the winter months, many of the low-pressure systems entering North America sweep through this region. Some of the lows bring strong gale-force winds and sometimes higher, driving rain almost horizontally. "We get storms here that would devastate the East Coast," said Mike Saindon. "The rain comes in three different directions at the same time, with wind gusts up to 100 miles an hour." The heavy storms cause many power outages. Tree root systems in the Pacific Northwest tend to be shallow, and heavy winds, combined with wet soil, cause trees to topple onto power lines.[12] During the winter months it is not unusual to have long blackouts.

It is during the winter months that the Pacific Ocean does not live up to its name. There are no barrier islands between the Northwest coast and one of the longest unbroken stretches of ocean in the world. The ocean bottom tends to shallow in a steep incline at a short distance from the beach. Both of these factors, combined with strong westerly winds, cause waves high enough that 20-foot surf is not looked upon as out of the ordinary. When a strong wind brings high surf and it combines with high tides, the wildly surging water can take large driftwood logs more than 2 feet in diameter and 70 feet long and throw them about like matchsticks. At these times, the beach can be a very dangerous place, as two crew members from the Quillayute River station would find in the predawn hours of 12 February 1997.

The combination of isolation and periods of poor weather caused one former officer-in-charge of the Quillayute River station to dub the unit "the fortress of solitude."

2 Welcome Aboard

In 1996, I decided I would write a book about the small-boat rescue stations of the U.S. Coast Guard. Thirty-eight years earlier, my first duty station in the U.S. Coast Guard was aboard a station in, of all places, Indiana, where it touches Lake Michigan at Michigan City. Obviously much had changed at the units during the preceding three decades. So that readers could better understand the lives of the men and women at the stations, I made arrangements to visit a number of stations throughout the United States and to live aboard the units. On Tuesday morning, 11 February 1997, I departed from home for my visit to the Quillayute River station.

There has been a U.S. Coast Guard station at Quillayute River since 1932. Originally sited on a hill that overlooked the entrance to the harbor, the unit was relocated approximately one-half mile upriver and completely rebuilt in 1980. After the 1997 deaths of the three U.S. Coast Guardsmen, there would be many "what-ifs," or talk of "fate," and the location of the station is the first of the what-ifs. If the station had been in the original location, the what-ifs state, there is a chance the officer of the day could have checked immediately to see if the sailboat *Gale Runner* was, in fact, on the bar. (A bar is a bank or shoal, usually at the mouth of a river, which makes entry into the river difficult or impossible.) Furthermore, the reasoning continues, Master Chief Boatswain's Mate (BMCM) George A. LaForge, the officer-in-charge, might have been able to observe quickly whether the motor lifeboat CG 44363 was being set against James Island.[1]

The single-story brick structure contains offices, a messing facility, and barracks. The main entrance is through double glass doors near the visitor's parking area in front of the building. The foyer serves as the quarterdeck, a place to greet people. The first thing that catches your attention is a large colorful mural that depicts a 44-foot motor lifeboat from the station towing a fishing vessel. One finds at all small-boat stations an amazing amount of talent: the large mural was completed by a crew member, Terry Michalski, in 1981, with the same paint used to paint the station boats.[2] Immediately to the right in the foyer is the communications center, with a window and sliding glass so the person on watch can talk to visitors. The communications watchstander greeted me and I identified myself. He told me to please wait while he informed the executive petty officer.

In a few minutes Boatswain's Mate First Class (BM1) Jonathan "Jon" Placido came out to greet me. He ushered me through the door to the right off the foyer. Through the door, a work area to the left, known as the operations area, has a desk, a computer, a status board, and other office equipment. To the right is the communications room. A very short hallway leads to the offices of the executive petty officer (XPO) and the officer-in-charge. The offices are connected to each other by a door.

Jon told me that Master Chief LaForge would be gone most of the day, as he had business in Port Angeles. Jon had just returned from medical attention at Group Port Angeles for a pain in his back. An earlier search and rescue case had caused the injury.

In 1997, BM1 Jon Placido had 11 years of U.S. Coast Guard service. In addition to being the XPO, Jon held one of the three surfmen billets at the station. "Ironically, I was born on an Indian reservation at Hoopa, California, but consider myself from Brookings, Oregon," Jon said. He entered the U.S. Coast Guard in 1981 because of the recession at the time.

"I was originally going to go into the marine corps, but me and a couple buddies of mine met the marine corps recruiter down in Eureka

[California]. He thought the way to get us in was to take us out and have a good time. He took us into this pizza place, bought us a bunch of beer. He got into a fight and got hauled off to jail. So, I didn't join the marines.

"I had some friends who were in the Coast Guard stationed on the *Cape Carter* [a 95-foot patrol boat] down at Crescent City [California]. I was working a full-time job and trying to go to school. My friends were always going out and having fun. They always had this expendable income. I was trying to make car payments, house payments, trying to make everything meet and trying to go to school. I never had an expendable income. I thought, '*Goddamn,* this is great!' I guess I came in for the expendable income: Not having to worry about rent, utilities, things like that.

"I was in almost six years. I got out because I was getting ready to get married and wanted my freedom. I was working in construction and doing really well. In some cases I was earning close to 20 dollars an hour, but the bottom line was that as I looked ahead, and I started to mature, the only thing I could see [was] I would work until I was too old to work anymore. I didn't see a light at the end of the tunnel.

"I wanted that freedom at the end of the tunnel. It's not much, but the U.S. Coast Guard retirement will give me what I want. I would love to have a piece of property out in eastern Oregon someplace. Build my own house on it. Have my family and a dog, a pond where I could hunt ducks. I don't have these extravagant dreams of having yachts and touring the world. The retirement pay will sure make that house payment."

In 1997, Jon had eight years of experience at small-boat stations on the Oregon and Washington coast, with three years of service aboard a buoy tender. When he came into the U.S. Coast Guard, he did not really care where or what type of duty he got, but he did want to go to Alaska. As it happened, his first duty station was a buoy tender, the *Planetree,* out of Juneau, Alaska.

Jon is six feet tall. He has the bluff, no-nonsense exterior of an XPO. Anyone who has ever held the position of executive officer at any command in the armed forces will tell you that the job is not fun-filled. Many times he or she must support and enforce edicts they may not personally agree with. As a captain once said, "It is fun being the king; it is not always fun being the assistant to the king." Then there are the personnel matters. Those in command positions at small-boat rescue stations will tell you most of their time is taken up with people matters.

Jon said that when he took over as XPO, the former XPO "had been laid-back and Senior Chief [Dan] Shipman [the former officer-in-charge] was laid-back. I had to be the hard-nosed guy so Senior Chief Shipman could be the good guy. The same with Master Chief LaForge."

I would get to know Jon better over the months to come, but my first impression told me he expected his crew members to perform to the best of their capabilities. Seaman (SN) Jacques Faur recalled that Placido "worked us hard, he always demanded nothing but 100 percent of you." Others would echo Faur's "he was a professional." Jon Placido's professionalism would be severely tested within 15 hours of my talking with him in his office.

Fireman Apprentice (FA) John D. DeMello recalled a case after the February 1997 deaths when the station had received a "Mayday"—the international call for distress—from a sailboat. "Petty Officer Miterko had already gone out on another Mayday call. There were two different calls close together. He's telling us, 'Watch out for this, watch out for that. It's really breaking bad, it's really big out here and nasty.'

"So, we're getting ready to go across the bar. We're sittin' in the river and the boat's already rocking bad and I'm thinkin,' 'We're still in the river. It's gonna be bad out there!'

"The XPO said, 'Look, I know you guys are going to be afraid, it's getting kind of bad out here. I'm scared too. But we got a job to do. There's a sailboat that's going to crash. We can do it by all working together.'

"We went from, like, hanging our heads, to bringing our heads up and thinking, 'He's right. This is our job and we gotta do it.'

"So, we went across the bar and everything was okay. I was afraid the whole time. It was so bad that night that when we met the sailboat at Sea Lion Rocks, and we turned around—we were going with the waves out to Sea Lion—we turned around and started battling the waves. And we were going for 20 minutes. I remember looking at my watch and wondering how long before we get home. We went 20 minutes and I started to get real sick and the XPO said, 'Look, you want me to turn around?' So, he went with the waves and I started to feel better and I puked and stuff. [Placido made the turn down swell to give DeMello a better ride.] And then we went the other way for five minutes and I turned around and looked right off my port side and there was Sea Lion Rocks. I said, 'How'd we travel for 20 minutes one way and turn around for five minutes and still be at the same place we started off at?' The XPO said, 'That's how it is out tonight.'

"The XPO kept keeping our hopes up that night. He knew we were afraid. He knew *I* was afraid. I had told him straight out I was afraid to get on the boats after the February incident when the motor lifeboat capsized. He said, 'Come on, we're all afraid.' Whenever he seen doubt, he tried to liven things up with a joke, or saying something sarcastic, or making you feel bad about feeling scared. Just to prove him wrong, you'd perk up."

Seaman Apprentice (SA) Falicia "Fe" Brantley said, "XPO was awesome. He was mean, as mean as they get, but if you worked hard, he was cool. He'd back you up. He always did."

BM1 Placido earned his surfman's qualifications in the demanding Pacific Northwest. Within the tight small-boat community and the tighter surfmen group, there are legendary stations and officers-in-charge. Jon served at one of them: Depoe Bay, Oregon, to the south of Quillayute River about half way down the Oregon coast (see map 1 in chapter 1). This unit is hellish. To reach the safety of the harbor you must run through what seems a very small 50-foot opening in a rock cliff. (The crew has been called the "hole in the wall gang.") Imagine going through this opening in heavy wind and seas, towing a disabled boat. One surfman said, "You get used to working closely around rocks." Jon broke in under Chief Boatswain's Mate Wayne Marshall, the officer-in-charge. Another surfman told me, "When I was getting my check ride for coxswain, Chief Marshall was doing my evaluation. I was towing a boat through the hole. The chief looked me in the eyes, smiled, said, 'Man overboard!' and jumped over the side!" Normally, a man-overboard drill is done with a dummy called "Oscar" and is hard enough to do with a boat in tow in a narrow location; try to imagine how it would be to do with a real person in the water, and that real person your officer-in-charge.

"I specifically asked not to be stationed at Quillayute River," Jon told me. "I asked to go to any floating unit. The detailer [officer at headquarters in charge of making personnel assignments] told me he had two 378s [378-foot-high endurance cutters] he could send me to, or he could send me to Quillayute River as the XPO. I had to have sea time to make chief. So I told him, 'Let me check into those two 378s and see what one I want to go to.' He said, 'I want to hear from you tomorrow.'

"The next morning I had orders to Quillayute River. Mike Saindon and I had qualified as surfman at the same time, so we had an extra surfman at Depoe Bay. Senior Chief Shipman only had one surfman at Quillayute River and himself. I went up there TAD [temporary additional duty] during December, January, and February as the third surfman.

"I will admit, when I sent in my dream sheet [form on which people list stations where they would like to serve], I said that I needed a sea billet to make chief, but I would like to be considered for the Quillayute River job, provided I couldn't get a sea job. The detailer heard: 'Send me to Quillayute River, please.'"

Jon pointed out that three small-boat stations are so isolated that no one wants them: Quillayute River and Neah Bay, Washington, and Grand Isle, Louisiana, and if "you even think of one of those stations, you are gone."

Jon felt the largest difference between Quillayute and the other small-boat stations he had served at had nothing to do with boat-handling or sea conditions, but with the community he served. At Chetcho River, Oregon, his first small-boat station, Jon's family had lived in the area since the early 1900s. At Depoe Bay, he explained, the community is noted for the support it gives the station. At Quillayute River, he found the attitude toward the unit "indifferent." At Depoe Bay, the command did not have to work to get people to want to serve the community; it was the opposite at Quillayute River.

Early in my conversations with Jon Placido, I thought he projected the self-confidence—some call it the macho—image of a D13 [Coast Guard District Thirteen] surfman. This image requires some explanation.

3 Boat Drivers

Most Americans feel that when a search and rescue (SAR) alarm rings at a small-boat station, anyone stationed at the unit can jump aboard a boat and make a rescue. That perception is wrong. The U.S. Coast Guard has an intensive program to train the people who crew their boats. The person in charge of a boat must go through a long training period and accomplish certain requirements before being allowed to pilot a station's boats. The training is a combination of on-the-job work aboard a boat and studying manuals. The first step in the process of becoming a coxswain—the person who navigates the boat and is in charge—is to serve as boat crew member.

The time needed to qualify for boat crew depends largely on the person and how much time he or she can get under way. BM1 Mike Saindon said, "A 'normal' length of time at Quillayute River is approximately three months to qualify as a boat crewman and then another four to eight months to qualify as a coxswain. In some cases it may take longer." To receive the qualification, the person must pass a board made up of a qualified surfman, boat coxswains, and engineers.

The atmosphere at a board reminds one of the oral examinations for a Ph.D., with the person being questioned as tense as a doctoral candidate. At one board a fireman apprentice in the engineering department fielded questions such as, "Can you name the controls of a 44-footer?" "How many fire extinguishers on the boat? What type and weight?" A petty officer then asked, "What type of aid to navigation is number 2, and then tell me all the aids upriver from that point." The fireman apprentice had to know 11 miles of river. The questioning continued for an hour and a half, after which the man left the room while the petty officers deliberated.

The members of the board could pass the fireman apprentice, fail him, or have him study certain items and be tested just on these items. In this case, the board decided the Coast Guardsman was deficient in only a few areas and would not have to sit through another grilling by a board, but rather had to remain at the station until he could successfully pass an examination put to him by a petty officer.

Once qualified as a crewman, a person can become either a coxswain or a boat engineer. To become a coxswain, a person must be a boatswain's mate or boatswain's mate striker (a person who gains the rate by on-the-job training). A crewman becomes a boatswain's mate by requesting the officer-in-charge, or commanding officer, to enter the rating; if approved, the crewman begins to take correspondence courses, learn from senior petty officers, or attend service schools. By the time a person makes it to this plateau, he or she is usually a boatswain's mate third class. Most coxswains have the boatswain's mate rating, but coxswains can sometimes have other ratings.

The U.S. Coast Guard gives an enlisted coxswain a great deal of authority when under way in any small boat. The service says the coxswain "has authority and responsibility which are independent of rank and seniority in relation to other personnel embarked." In other words, no matter what the rank of the person on the boat, when under way on a mission, the coxswain is in complete charge. BM1 Bart Pope, of the National Motor Lifeboat School at Cape Disappointment, Washington, told me of a time he had two visiting captains from the district aboard a motor lifeboat when fog rolled in. "I put the captains on watch and they reported as ordered."

An idea of what a coxswain must know can be gained by glancing at the training workbook and reading the requirements to pass the test for boat driver. Task COX-06-01 is the recovery of an actual person from the water "using the direct pickup method." The prospective coxswain has to perform this task "under way during daylight hours in fair weather conditions with calm seas. . . . Trainee must accomplish task without prompting or use of a reference." The student coxswain is judged on 11 items, with a standard that states: "Task must be done without placing the person in the water in any type of jeopardy what so ever. Boat's engine(s) must be in neutral when the PIW [person in the water] is brought alongside. Task should be performed so that pickup is within two minutes of the time when the first warning is given. To show competence the trainee should be able to do the task on the first attempt. Extensive maneuvering would indicate a lack of the necessary mastery." For recreational boaters who think they might be able to do such a task on a calm day, there is also a task to do a recovery with "a life-like dummy . . . in 6 to 8 foot seas." The standards section of this task says, in part, "Task should be performed so that pickup is within three minutes of the time when the first warning is given. To show competence the trainee should be able to do the task on the first attempt."[1]

The coxswain must also pass a board and a check ride usually conducted by the executive officer or the executive petty officer of the unit. A letter of qualification is placed in the new coxswain's service record. The person may lose the qualification—"have the letter pulled"—for serious breaches of conduct. The coxswain must be recertified every six months. If transferred, the person must be recertified at the new station. According to BM1 Saindon, "It usually takes a person six to eight months after making boat crew qualifications to become a coxswain at Quillayute River."

The only person on a boat who may relieve the coxswain is the coxswain's commanding officer, officer-in-charge, executive officer, or executive petty officer. A senior officer in charge at the scene of a distress can also relieve the coxswain.[2]

The authority given to a coxswain by the U.S. Coast Guard is also tempered with restrictions on what types of conditions the person can operate within. In February 1997, the restrictions put upon coxswains stated that a coxswain *could not operate a boat on a breaking bar.* Only a surfman was considered qualified enough to take a boat out in those conditions.

The ultimate qualification for a small-boat operator is that of surfman. Until recently, most surfmen were in the Thirteenth Coast Guard District, although this would change after 12 February 1997. The training record book for surfman is 28 pages, with 18 pages devoted to instructor sign-off sheets of required tasks. Task SRF-01-10, for example, details the requirement for demonstrating ability to recover a simulated person in the water in six- to eight-foot surf. Among other standards for this task is one stating that the "boat must proceed from a point at least 150 feet seaward from the simulated person in the water and accomplish a recovery." Recall that this is in six- to eight-foot surf, which is a more dangerous environment than regular seas. The trainee is judged on 15 performance criteria.[3]

At the Cape Disappointment, Washington, station, I observed the surfman boards for two petty officers. The board consisted of the senior people at the Cape Disappointment station: Lieutenant Michael "Mike" White, the commanding officer; Senior Chief Boatswain's Mate Thomas "Tom" Doucette, the executive petty officer; Senior Chief Machinery Technician Michael "Mike" Doan, the engineering petty officer; and BM1 Bart Pope (this was prior to his transfer to the National Motor Lifeboat School). The questioning came mainly from Senior Chief Doucette and BM1 Pope, both surfmen. Most of the questions dealt with equipment and conditions, plus decision-making. "What would you do if you were in the 52-foot motor lifeboat and the windows blew out?" "How would you quickly drain the water out of the compartment?"

I especially recall a question to BM1 Chris Smashe: "What should you do on a rollover?" Smashe repeated the material from the book. Doucette added: "Remember to cut back your throttle. The faster your engines are turning, the more you will flood your engine room. A rollover with screws turning can kill a crew member if he should go back along the keel. You can survive an out-of-gear idling screw."

The responsibilities of a surfman are great. The service specifically mandates that a boat cannot be operated unless a surfman is present on the boat in certain weather and surf conditions. Further, the surfman becomes an expert advisor to the officer of the day and commanding officer during heavy weather conditions.

A surfman, however, told me that the real date he became a surfman was not the date on the official letter of certification, but "when I got the crap scared out of me on the bar climbing up the face of a 20-plus-foot breaker, falling 20-plus-feet off the back side, and then pulling myself off the window just in time to meet its big brother. After that was over, it was time to do it all over again, but looking over my shoulder on the inbound."

Beginning in 1996, a surfman earned an extra $220 each month in special pay. As of November 1997, surfmen also were allowed to wear an insignia on their uniform that is a life ring with crossed boat oars—the original logo of the U.S. Life-Saving Service. A letter of qualification is placed in the surfman's service record; the surfman must be recertified every six months and must be recertified if transferred to another station.

In any of the other armed forces, the surfmen would be considered an elite group. The leadership of the U.S. Coast Guard, however, seems to discourage this type of thinking. Indeed, there seems to be a movement on the part of some decision-makers to rid the service of the small-boat stations (see part 3). Most of the people "in the field" can sense this attitude and therefore direct strong animosity toward the officer corps. Retired Chief Boatswain's Mate Steve Huffsteadler said, "These days it makes me more sad than anything else that a proud segment of the Service [the small-boat rescue stations] is of so little value to its leadership. That the job I spent 20-plus years doing has so little meaning to anyone but us lifeboatmen and to the people who we saved. I know that I didn't waste those years and I know what I did had value, but I fear that as time goes on, I will be the only one that knows."

The surfmen recognize this attitude of the leadership. They also recognize that the environment of the Thirteenth Coast Guard District is unique and have developed what some officers call an "attitude" or have become too "macho." The surfmen, however, are simply a group of people who know they have reached the height of their profession and can operate in some of the worst conditions facing those who put out to sea in small rescue boats. In fact, because of the consistently high surf in the Pacific Northwest, those who advance to coxswain in the Thirteenth District feel they are better prepared for heavy weather than their counterparts in other districts.

4 Port Duty Section

After a short talk with BM1 Jon Placido, he informed me I would be bunking in the transit quarters while at the station. He escorted me out of his office, down the hallway, out into the foyer, and through the door on the left-hand side of the entranceway that leads to a small recreation ("rec") deck and the mess deck. We went out a set of double doors off the rec deck, into a breezeway where the fire-fighting gear hung, and through another doorway into the barracks.

One of the changes between the old Guard and the new is that the U.S. Coast Guard now provides for government or government-leased housing away from the station. The station barracks is made up of individual rooms, one for the duty officer of the day (OOD) and surfman. Usually, only those with the duty, or those who do not yet have quarters, stay aboard. When living aboard, a crew member usually shares a room with at least one other person, if not more. Because of the isolation, most people assigned to the station who have government-leased housing live in Forks, about a 20-minute drive from the station, although there are a few government quarters less than a mile from the station. The officer-in-charge lived in one of the units close to the station.

After showing me my room, Jon took me out the back door of the barracks and guided me through the engineering spaces in a building to the right. Immediately behind the barracks is something not seen at most small-boat stations: a small gym. Most stations lack amenities for their crews. Because of their isolation, the Quillayute River station and the station at Neah Bay, Washington, Quillayute River's neighbor station to the north, are the only small-boat rescue stations I have seen with a gym.

After stowing my gear, I went back to Jon's office. Jon took me to the operations work area near the communications room and said, "This is Boatswain's Mate Second Class (BM2) Bosley, the OOD. He will take care of you now." Jon returned to his paperwork.

On 11 February 1997, 24 people were stationed at Quillayute River. The crew was organized into a work schedule that included those who stood senior duty officer and surfman duties; those who stood OOD watches; those who were assigned to duty sections; and those who were known as day workers.

BMCM George A. LaForge, the officer-in-charge, and BM1 Jon Placido stood both senior duty officer and surfman duties. BM2 W. Brent Cookingham, the other qualified surfman on the station, was operations petty officer and stood surfman duties. The station's complement called for four surfmen, but there were only three aboard. Because of the shortage, Master Chief LaForge would usually be the surfman for the standby boat.

In February 1997, U.S. Coast Guard regulations stated that if there were adverse weather conditions in a station's area of responsibility, a surfman had to be aboard the boat. The Quillayute River station's operating instructions, however, stated that *if the weather was favorable, the surfman was allowed to go home, subject to recall.* Because of the shortage of surfmen at Quillayute River, if it were not for this recall regulation a surfman would have had very little time away from the station. Every day the duty surfman drove the government pickup to an observation point near the old station, called "the hill," to check the bar. The observations were taken early in the morning, at noon, and before sunset. The frequency increased during threatening weather. The surfman also checked with the OOD about the weather. *The senior duty officer and surfman rely upon the OOD to pass correct and timely information to them,* an expectation at all stations and ships.

The Quillayute River station's engineering petty officer, Machinery

Technician First Class (MK1) Mumford, did not fall into the regular watch rotation, but did fill in when needed. Mumford would later state that "because of personnel shortages" he often stood duty.

Considered "day workers" were the two cooks at Quillayute River and Damage Controlman Second Class Michael Keller, who worked primarily on the U.S. Coast Guard housing units.

Three boatswain's mates were qualified as coxswain: BM2 Bosley, BM3 Donald J. Miterko, and BM3 Paul R. Lassila. Miterko was in training for surfman. The OOD duty rotated between these three men.

The rest of the people aboard the Quillayute River station were divided into two duty sections, "port" and "starboard." Each duty section stood two days of duty during the week and had every other weekend off. Duty days began at 1:00 P.M. Port section had just had their weekend duty and had started their first duty day for the week on Tuesday, 11 February. The section had seven people in it, with three petty officers: Machinery Technician Second Class (MK2) Thomas L. Byrd, BM3 Marcus M. Martin, and Machinery Technician Third Class (MK3) Matthew E. Schlimme. The other people in the section were Seaman John A. Stoudenmire III, Seaman Clinton P. Miniken, Seaman Apprentice Benjamin F. Wingo, and Fireman Apprentice John D. DeMello. BM3 Martin was a break-in coxswain.

On 11 February 1997, BM2 Bosley would begin his OOD duties. He would also be the ready boat coxswain, the first boat out in an emergency. His crew would be MK3 Schlimme, Seaman Miniken, and Seaman Apprentice Wingo. Wingo had been at the station for only two months and was a break-in crewman.

The backup, or second boat, would have a crew of MK2 Byrd, BM3 Martin, and Seaman Stoudenmire, with BMCM LaForge as the backup surfman. If weather conditions and work permit, those on the second boat are allowed to go home, subject to immediate recall. If the ready boat should get under way, then the second boat crew is automatically recalled to the station.

BM2 Bosley's assignment as OOD on 11 February was an important one and was critical to the events that would take place less than 12 hours after he assumed his duties. The OOD is the first-line petty officer who will make the initial decisions concerning a case. If the officer-in-charge and XPO are aboard, the OOD will inform them of what is happening. When neither senior person is aboard the station, and when a surfman is not aboard, the OOD is in charge of the unit.

BM2 David A. "Bos" Bosley, 36, of Coronado, California, had come into the U.S. Coast Guard after serving in the U.S. Marine Corps, to "drive boats and save people." Bos had transferred from the Yaquina Bay small-boat rescue station, located in Newport, Oregon, in 1996. Bosley was "always 'a water person.' He loved the water, the ocean, swimming, surfing and fly fishing." BM2 Frank Hiebert, who served with Bosley, would recall the boatswain's mate's love of fishing and pointed out that fishing was also "an excuse to cruise the logging roads and to get away from the station."

Bosley had, according to Hiebert, a "type A personality," but Frank recalled a time when he had broken up with his girl and "for weeks we were avoiding each other. One evening Bos stormed into my room, told me to get up and get dressed, that I was going to dinner! He had paid in advance for our dinner at The Smokehouse, and was gonna babysit Tina's four-year-old daughter. We went, and decided to get married. This is the side of him that made me call him friend."

Four years previously, Bosley had won his first major U.S. Coast Guard award for a rescue at Newport. On 7 March 1993, Mark Nelson, his son, James, 10, and Nelson's sister were returning from an aborted fishing trip when a breaker at least 16 feet high capsized their 18-foot boat.

"One of the guys on the jetty saw it capsize," Nelson would later recall. "He yelled at someone with a cell phone, and he called 911."[1]

Chief Boatswain's Mate Glen Butler Jr., now retired, came out in a 30-foot surf rescue boat (SRB) with Bosley and another Coast Guardsman aboard. The SRB is a very fast boat designed to run quickly into the surf zone, pluck someone from the water, and transit rapidly out. Chief Butler soon had the SRB maneuvered alongside the three people clinging to their overturned boat in the cold 52-degree sea water. Then another breaker slapped the SRB and drove it away from the Nelsons. The breaker had driven the three people in the water farther out into the choppy sea. Chief Butler maneuvered the boat again toward the people. The youngest, James, was the first rescued.

Bosley, who was tethered to the SRB with a lifeline, jumped into the water and swam toward James. James's father much later would recall, "I remember him [Bosley] calling: 'James, I'm coming for you! Stay calm.'"

Bosley managed to reach James, who had on a life jacket. Bosley's lifeline, however, had become fouled. The waves were battering the SRB, dragging Bosley and the boy backward. Bosley released James, turned, and cut the lifeline. He then swam back to James and brought him alongside the SRB. Twenty-one minutes after the U.S. Coast Guard received the call, all three people from the fishing boat were rescued. Bosley later stopped by the hospital where the three were being treated for hypothermia. For his work, Bosley received the U.S. Coast Guard Commendation Medal.

Chief Butler would later say of Bosley, "He never turned himself away from an SAR [search and rescue]. That guy was always ready to go. You could count on him."

Some crew members, however, felt that Bosley might be a little too anxious to get under way. One crew member summed it up: "Bosley was a good coxswain, confident boat driver, cared about his crew, and looked out for others. However, he had a tendency to be aggressive and pushy about getting under way fast without regard to proper preparation. He would get upset if someone took the time to get suited up properly, and he didn't always brief his crew. . . . Bosley had a cowboy-like attitude."

Fe Brantley later would say, "I couldn't stand Bosley when I first got to Quillayute River. I just hated him. Then one day I was walking down the brow to the boat. I had a whole bunch of stuff in my arms for painting the boat. I was overloaded. They used to tell us not to walk down that brow overloaded, or you're going to fall, or something.

"I dropped the jitterbug [a tool for removing paint] over the side. You know how when you've got to go tell your parents something, you walk real slow. I was walking real slow to the phone. I called up Bos. I said, 'You know what I did? I dropped the jitterbug into the water.'

"He said, 'Did you? Well, go get another one. Hey, Fe, shit happens.'

"After that, instead of being a little smart ass to him, I started trying to listen a little bit to him. I think with me trying to listen to him, it made him treat me differently."

Sandi Bosley, Dave's wife, would later say that Bosley came into the U.S. Coast Guard from the marine corps "because he'd rather save lives." She also said, "he loved driving boats."

Bosley introduced himself to me and then said he would detail BM3 Paul Lassila to show me the station's boats.

5 Motor Lifeboats and Surf

BM3 Paul Lassila and I went out the front door of the station and started toward the boathouse. At the Quillayute River station, the boathouse is located about 100 to 200 yards west of the unit, across the reservation's street. A tourist would pay for the view from the station's front steps. Looking seaward, the high, steep cliffs of James Island dominate, at 183 feet in elevation. An elevated power line stretches from the mainland to James Island, with some aids to navigation on the island (see map 4 in chapter 18).[1] James Island is an important feature for the Quileute. Prior to their contacts with Europeans, the island provided a defensive bastion against any enemies. It is also the location where the tribe buried their chiefs. To the right is Little James Island, whose rocks tower 162 feet into the air, and the next piece of rocky land is Gunsight Rock. Little James Island and Gunsight Rock line up like soldiers marching out to sea.

The view to the left of the station entrance reveals a boat basin, with a small number of commercial fishing craft. Also to the left, one can see the Quillayute River and a few glimpses of the ocean.

Paul and I walked along the pier leading out to the boathouse. The Quillayute River station has a small boathouse, with room for only one of the unit's two motor lifeboats. The station does not have a haul-out facility, so any work that entails taking the boats out of the water must be accomplished at another location.

The two most important items at a small-boat rescue station are the people and the boats. On 11 February 1997, the primary boats at the Quillayute River station were two 44-foot motor lifeboats, CG 44363 and CG 44393.[2]

The motor lifeboats in use in the U.S. Coast Guard can trace their development back to England. In 1873, the Treasury Department—under which the U.S. Life-Saving Service fell—brought into this country an oar-powered, two- to four-ton lifeboat for testing. Several of the surfmen of the U.S. Life-Saving Service did not like the boat because of its great weight. The boats normally used, called surfboats, were light and more maneuverable, and only required six men. The ponderous lifeboat required at least eight men. Then an interesting thing happened. Reports from the Great Lakes began to filter through the service that the boat could take almost impossible seas. Skepticism changed to admiration, and crews began to regard the lifeboat as "something almost supernatural," for it enabled them to provide help "when the most powerful tugs and steam-craft refused to go out of the harbor." Thus began what could be called a love affair between those who push out into high seas and the boats they depend upon.

The early lifeboats were self-bailing and self-righting. Just because a boat will right itself does not mean there is no longer any danger. Shortly before sunrise on 23 April 1880, Keeper (the man in charge) Jerome G. Kiah, of the U.S. Life-Saving Station at Point aux Barques, Michigan, and his crew of six surfmen shoved off into a heavy storm on Lake Huron. The boat was capsized many times and the crew managed to right it each time, but eventually only one man, Kiah, survived the cold waters of Lake Huron.

In 1899, the U.S. Life-Saving Service began evaluating the results of putting a 12-horsepower gasoline engine in a 34-foot lifeboat at their station at Marquette, Michigan. When the U.S. Life-Saving Service and U.S. Revenue Cutter Service merged to form the U.S. Coast Guard in 1915, 80 motorized lifeboats were in service.

The new U.S. Coast Guard introduced a 36-foot motor lifeboat in 1918, the Type H. Built in the U.S. Coast Guard Yard, Curtis Bay, Maryland, the boat underwent four major design changes between the two world wars and became the standard heavy-weather boat in the service. One can still see the basic shape of the old English lifeboat of 1873 in the lines of the new boat. The boat continues the self-bailing and self-righting features. A lookout position is located in the bow. The forward portion of the boat contains an enclosed cabin for survivors, and the engine is in a compartment in the center of the boat. The crew stands in a lowered cockpit near the stern of the boat, and a rear compartment holds the towing line. A towing bit is located just aft of where the crew stands. The boat has a single screw and rudder. There is nothing fancy about the craft; you stood out in the open and took whatever the sea and the weather threw at you. As Master Chief Boatswain's Mate Thomas "Tom" McAdams, a retired legend in the small-boat community on the West Coast, once observed, you "could not stay dry or warm on wet, cold cases." Electronics consisted of a radio and, perhaps, a fathometer.

The new motor lifeboat proved a reliable boat that could take an astonishing amount of punishment and still bring back its crew. The 36-footer could make maybe 8 knots on a good day, but it might also make 8 knots on a very rough day. It was like the tortoise versus the hare. Many a sleek pleasure craft could run circles around the motor lifeboat, but, when the seas began to break, the plodding lifeboat could continue the course while the flashy boats ran for cover. Generations of U.S. Coast Guardsmen swore by the 36-footer, including the author.

In 1961, the U.S. Coast Guard built, again at the U.S. Coast Guard Yard, a prototype 44-footer. The service delivered CG 44300 to the West Coast for testing in the high, powerful surf of the Pacific Ocean. Chosen to conduct the test was Chief Boatswain's Mate Giles Vanderhoof. Chief Vanderhoof subjected the boat to *35-foot seas.* He deliberately broached the boat, an almost fatal maneuver for most boats in heavy seas. Many times he brought the boat back to the engineers to repair, while picking glass out of his wet suit from when the windows blew out. At the end of five days, the main cabin was crushed and holed, the aft cabin was smashed, the screws and struts had been torn loose from the hull, the engine mounts had given way, and both engines were lying askew in an engine room smeared with oil. But *the boat was still afloat.*

The strength of the small 44-footer can be grasped by looking at the history of this first boat. In 1995, the CG 44300 was still afloat even though it had rolled over six times, pitchpoled (gone end over end) three times, and been rammed by a freighter while in its boathouse. Yet it kept its crews alive. A total of 110 44-footers were built. In 1995, 99 remained in service. Old CG 44300 finally reached the end of her life. The boat, assigned to the National Motor Lifeboat School, located near the dangerous Columbia River bar, had been placed as a backup boat for the Cape Disappointment station and blew an engine while helping in a search and rescue case. In 1997, the boat was put on loan to the Columbia River Maritime Museum, Astoria, Oregon, where it will soon be a part of a major new U.S. Coast Guard exhibit. In September 1990, the U.S. Coast Guard began testing a prototype 47-foot motor lifeboat.

The 44-footer is so different from the 36-footer that old lifeboatmen have said that it was like going from an old pickup to a Ferrari sports car. The 44-foot motor lifeboat has a maximum speed of 15.3 knots, with a cruising speed of 11 knots and a range of 200 nautical miles at 10 knots and 150 nautical miles at 15 knots. The boat is powered by two GM 6V53 diesels, with 186 horsepower at 2,000 rpms. It is twin-screwed and has twin rudders. The boat has a displacement of 36,000 pounds and like its predecessors is self-bailing and self-righting.[3]

The 44-footer has the coxswain sitting in a chair forward in the boat, 20.5 feet off the water. The area where the crew stands is called the coxswain flat. The four-person crew is now protected by half-inch-thick, heat-treated glass, so they are shielded from the direct blasts of sea and weather. You will, however, still get wet on a motor lifeboat. In high seas the water hits the windshield, swirls around it, and enters the coxswain flat. Like all lifeboats, the 44-footer rolls and pitches heavily, throwing spray and water upon the crew.

From the chair at centerline, the coxswain—the person navigating the boat and in charge—has a good sweeping view forward. Most of what a coxswain will need to run the boat is easily handled from the seat, or can be relayed quickly to a crew member. The coxswain, reaching up and to the right, can grasp the microphone for the radio; also close to the right hand are the two throttles, while further to the right is the radar. The radar must be handled by a crew member because it is out of reach of the person in the chair.

A few feet behind the coxswain chair is a wire mesh screen between the coxswain flat and the towing bit. Looking aft, the deck drops down to a well deck close to the water and then rises again to form the aft part of the boat. In this section is an area known as the survivor's compartment. The top part of the compartment is known as the aft turtle.

Climbing down from the coxswain chair and moving to the left side of the coxswain flat, you come upon a watertight door with a quick release. Through the door and down a few steps is the forward compartment, used as a galley and a resting place for the crew to get out of the weather during long rough cases. There is also a small portable head (toilet) that few people use, as the rule is "You use it, you clean it." Another hatch in the forward portion of the compartment leads to the engines.

The 44-footer will take an enormous amount of punishment. Chief Warrant Officer Two Thomas Doucette told me of an incident that happened near the Grays Harbor, Washington, station.[4] A 44-footer had a line fouled in its screws and was heading for a breaking bar. Crews are trained to remain with their boats. The crew, Tom related, went into the forward compartment, fastened down the door, and strapped into the seats with the safety belts provided. The boat rolled several times before help could arrive. The crew, albeit sick from their rollovers, survived the incident. The important thing to remember here is that the crew could get into one of the compartments and strap themselves in.

Boat crews usually consist of four people: the coxswain, a boat engineer, and two crew members. At times, there will be a break-in coxswain aboard. Because most stations have to operate short-handed, often one of the lookouts is also a break-in.

The coxswain is a petty officer, usually in the boatswain's mate rate. Responsible for safely navigating the boat and for everything that happens on the craft, the coxswain is often called upon to literally make life-and-death decisions in mere minutes, if not seconds, while trying to work a boat in almost impossible conditions. What has always amazed me is how young most of the coxswains are and how they make the tough, correct call, after which they are often subjected to the brunt of the unknowing media with little support from the senior officers in their own organization, as shown in the *Miss Renee* case.

On 14 November 1998, the fishing vessel *Miss Renee* wandered into the Clatsop Spit area, near the dangerous Columbia River bar, at night and in the teeth of a gale with driving rain. Clatsop Spit is noted for high, breaking surf. One area of Clatsop is nicknamed "death row" for the size of the surf in the area. The first call for assistance from the *Miss Renee* to the Cape Disappointment, Washington, station gave a location upriver, away from Clatsop Spit. The officer of the day (OOD), BM1 Stephen Schuch, rightfully dispatched the best boat available for a fast transit to the area, a 47-foot motor lifeboat. BM3 David M. Chapman II, coxswain of the 47-foot motor lifeboat, and BM1 Schuch soon learned the boat in trouble was not at the location given by the person aboard the *Miss Renee.* BM3 Chapman located the fishing vessel in Clatsop Spit and started toward the boat. Suddenly taking a series of 16-foot breaks (breaking waves), BM3 Chapman realized that this situation was above his abilities as a coxswain and, per regulations, moved out of Clatsop Spit.

The OOD, a surfman, quickly departed the station in a 52-foot motor lifeboat, but before departing recalled another surfman for the 47-footer. Aboard the 47-footer, the recalled surfman, BM1 Jeffery Kihlmire, located the capsized *Miss Renee.* BM1 Kihlmire, working in at least 16-foot breaks and shallowing water, and using outstanding boatmanship, pulled off a man clinging to the overturned hull of the fishing vessel. The crew of the 47-foot motor lifeboat then found that a teenage boy was trapped beneath the overturned vessel and that another crewmen was drifting away face down in the water. "My crew wanted to try to swim over to the boat and see if they could get the boy," Jeff Kihlmire said to me much later. "Dr. Noble, do you know how many Coasties have died trying to go under a hull to rescue someone?" BM1 Kihlmire made the hard call and did not allow his crew to make the attempt. Instead, he tried everything he could think of, but nothing would work. In the end, two people died.

On this gale-swept night two U.S. Coast Guardsmen made tough, but correct, decisions. A captain at U.S. Coast Guard headquarters told me much later that BM3 Chapman "had the balls to make a decision many captains in headquarters could not make." I would add that both BM3 David M. Chapman II and BM1 Jeffery Kihlmire made life-and-death decisions that most of the admirals at U.S. Coast Guard headquarters have never had to make. Yet the crew of the station, from the commanding officer down, were subjected to a media frenzy that laid the blame for the deaths on the Cape Disappointment station. What is even worse, there appeared to be no backup for this station. In short, higher decision-makers seemed to distance themselves from the controversy. Keep in mind this attitude by some senior leaders.[5]

The motor lifeboat crew has a boat engineer attached, who is usually a petty officer in the machinery technician (MK) rating. The boat engineer must also go through the crew member qualifications. Readers who are used to shipboard routines, or who have served in the U.S. Navy, may not realize how much people at small-boat stations are required to know outside of their field. On a motor lifeboat, the engineering rates must know such deck details as handling towlines and how to operate the radar. I once observed a MK3 learning how to radar navigate the motor lifeboat for her qualifications. A person in the engineer ratings usually takes three to five weeks to become a boat engineer after being boat-crew qualified, if the person is an MK3 when he or she arrives at the station. If a person is a nonrate in engineering, it will take three to six months after being boat-crew qualified.

The last two members of a normal motor lifeboat crew are made up of boat crew members. These are usually seaman and seaman apprentices, or firemen and fireman apprentices—the entry-level people. The two lookouts do the "grunt" work, handling lines both for mooring and unmooring, and for towing. If a person is to be recovered from the water, it will be one, or both, of these people who will either pull the people out or go into the water. They perform lookout duties and are additional eyes for the coxswain. Many a motor lifeboat coxswain has been spared the investigation into a rollover by a warning shout from a lookout.

Under way in a 44-foot motor lifeboat, the coxswain sits in the coxswain's chair. The boat engineer usually stands to the left; one lookout also stands to the engineer's left and on the steps to the well deck behind the engineer. The other lookout is to the coxswain's right.

When operating in cold waters such as those in Washington State, motor lifeboat crews are fitted with survival gear that includes thermal underwear, knit caps, boots, and gloves. All must wear an exposure suit. Most put on an orange exposure coverall that not only keeps them warm, but also provides flotation. Some crews now wear an exposure garment known as a dry suit. Over the top of their exposure gear, the motor lifeboat crew member dons a pyrotechnic mesh vest—called a pyro vest—with a number of pockets. Within the pockets are emergency signaling devices, such as a large day-and-night flare, a strobe light, and a small launcher for shooting the red-pencil flares that are also stowed in the pyro vest.

If the motor lifeboat crew is going to enter a dangerous area, or if seas are high, they are required to wear helmets to protect their heads. Goggles can be worn to help against the lash of spray. The crews are also required to wear surfbelts: a broad safety belt with two long straps attached that have swivel snaps at the end. The belt is clipped into "D" rings that are placed throughout the 44-foot motor lifeboat. The rings allow a crew member to move about the boat and yet always to be secured to the boat. Motor lifeboats, as mentioned, are designed to right themselves on a rollover. The "book" says that the boat will come up within 12 seconds of rolling over. Being attached to the boat, crew members will stay with their boat and not be thrown into the sea, which can be fatal in cold, high seas. At one time, the coxswain was belted into the chair, but this was found unsafe when a coxswain, tightly buckled in, was ejected from the boat when the chair tore loose. The surfbelt for the coxswain is now fastened to the boat. Just because a boat is designed to roll over, however, does not mean there is no danger, or that there will there be no damage to a boat. In almost all cases of rollovers, there is usually damage to a motor lifeboat. It is an unfortunate fact that what little media coverage there is about small-boat stations strongly hints that, because a motor lifeboat can right itself, there is no danger.

Steve Huffstutler, a former officer-in-charge of a small-boat station, said of the waters around the Quillayute River station, "That is probably the scariest piece of water I ever navigated, and I've sailed all of the Great Lakes, the Pacific, Atlantic, the St. Lawrence, the Panama Canal, you get the idea. Cape D [Disappointment] gets all the press, Depoe Bay [Oregon] is weirder, but for a sheer adrenaline rush, nothing beats the entrance to the Quillayute [River]. The bar at Quillayute River is strange and disorienting. Waves are deflected and reflected regularly." Tom Doucette, who served there, echoed Steve's observation and added, "At Quillayute River you have to keep tightly focused on crossing the bar. There are so many rocks in the area, if your attention wanders, you can be in serious trouble." The nature of the sea around the Quillayute River station is important to understand and helps to explain what happened on 12 February 1997.

Many of the harbors of the Pacific Northwest have bars near their entrances. Bars are areas of shoaling near the mouth of a river. Waves tend to break when the ocean bottom is steep, when a tidal current is against them, and in shoal waters. All of these factors come together in the Pacific Northwest to cause some of the most treacherous entrances to a port a mariner can attempt. At the Quillayute River station, when crossing the bar, coxswains are trained to steer 210 degrees from Wash Rock, just outside the entrance to the river, to the Quillayute River entrance buoy—known as the "Q" buoy.

Everyone can probably identify a wave, but, as K. Adlard Coles, author of *Heavy Weather Sailing,* has pointed out, "waves, like people, are of endless variety."[6] A sea wave begins by being generated by the wind, with its size dependent upon factors such as the strength of the wind, the time for which the wind blows, and the fetch—the length of the area over which the wind blows. Once leaving the generating, or fetch, area, the wave may gradually lose height. Waves that begin in the fetch area are known as wind waves, and wind waves that move away from the fetch area are known as swells. Swells are relatively easy to spot on a windless day, but once there is some wind and seas are forming they can be very difficult to recognize. The ways in which seas and swells react to obstacles, currents, and shoaling are important to the events of 12 February 1997.

If the direction of a wave is against a tidal current, usually in a shallower area, the wave steepens but its energy remains the same, so that high steep waves are produced. Waves "entering an area of opposing currents, can quite easily have their heights raised by 50 to 100 per cent in currents as low as 2 to 3 knots. Thus the breaking of waves may be a frequent occurrence even without much local wind."[7] Shallow water also has a considerable effect on a wave height.

When waves approach a coast, if they meet resistance such as an island, they can begin to bend. This bending is known as refraction. The wave nearest the island is slowed down, but the rest of the waves, in deeper water, retain their greater velocity until they reach shallow water. Islands can also cause another type of wave, known as a reflective wave. If a wave strikes an obstruction such as an island, it will bounce back, causing a confusing pattern. It can be difficult to judge the true wave direction when reflected waves cause different patterns in the area.

Breaking waves are extremely important to anyone on the sea. The oceanographer William G. Van Dorn writes, "Under extreme conditions, pressures in excess of one ton per square foot have been measured in breaking waves, while wind pressures rarely exceed 10–12 pounds per square foot."[8]

Another type of wave should also be mentioned: a freak, or rogue, wave. Laurence Draper, of the British Oceanographic Data Service, puts it well. "Amongst sailors there are many stories of freak waves, usually concerning great waves maybe 100 ft. high (which may or may not have grown with the telling) and which overwhelmed or badly damaged many a vessel, but to a yachtsman a wave 15 ft. high occurring on a day when the highest he's seen had been perhaps 8 or 10 ft. high is in every sense a freak."[9]

While many a landsman may think such stories of rogue waves are made up by salts to impress armchair sailors, incidents of these events can be documented by something other than a sailor's estimation. The *Daunt Light* Vessel was struck by a freak wave off Cork, Ireland, in 1969. "The Master described it as the most frightening experience in his sea-going life, and had it not been for the . . . fact that the wave recorder had been running, his story would probably have been put down to another bit of Irish blarney." The wave recorder showed sea conditions of 16.5 feet over a period of time; then suddenly, a 42-foot wave struck the vessel.[10]

There are many wave trains in the sea, each with its own period and height, traveling together at slightly different but constant speeds. The various components usually travel together in and out of step with each other. Then, every once in awhile, the waves get in step together at the same place and produce exceptionally high waves. "The life of such a wave is only a transient one, not much more than a minute or two in the deep ocean and even less in sheltered waters where the wave period is smaller."[11]

6 This Is a Drill

At 7:00 A.M. on 11 February, Fireman Apprentice John D. DeMello, 18, took over the communications watch at the Quillayute River station. He would have this duty for 24 hours until relieved at 7:00 A.M. on 12 February. John's duty is a critical one, and the training it requires is a good example of the amount of instruction the people at small-boat stations must undergo and of why a working week of 100 hours is not considered unusual.

The people at the U.S. Coast Guard Quillayute River station, as at other small-boat stations, must know how to be proficient with pistol, rifle, shotgun, and sometimes machine gun; know how to work in pollution control; be able to conduct search and seizures of illegal drugs; be able to conduct boardings for boating safety; know species of fish for fishery law enforcement; know federal fisheries laws in order to board fishing vessels to check for violations; know the procedures for boarding vessels to check for illegal immigrants—all in addition to learning the skills needed for boat crewman and coxswain: how to maintain a boat and its machinery; how to maintain a station's machinery; how to give first aid; how to fight fire; and how to search for lost boats. Furthermore, the crews must attend a certain number of lectures ranging from off-duty education to sexual harassment. This story of the Quillayute River station will focus only on the skills needed for search and rescue duty by the men and women of the unit.

A new person arriving at the Quillayute River station must first become qualified as a communications watchstander; this requirement includes cooks and those in the engineering rates, who on large cutters would not have this responsibility. The communications watchstander is on the front line of any case. Either telephone calls or radio calls will come to this person, so the watchstander is usually the first to sound the alarm. The communications watchstander at Quillayute River is responsible for handling the five telephone lines into the unit. In addition, the watchstander monitors the radio, which has a number of channels. Normally, all radio calls are received on channel 16, the international calling and emergency channel. If a boat calls and it is not an emergency, the caller will be shifted to a working frequency, usually channel 22A. The new person must learn about the "high site," the location where the antenna is placed at the highest location and the "low site," which at the Quillayute River station is the station. In 1997, the station's high site was on James Island. Knowing on which site a call is heard can help in determining where a call might originate. The person on watch must also know how to plot positions on a navigation chart, work a computer to receive messages, take and pass on weather observations to the group through a radio intercom, know all of the aids to navigation within the area of operations of Quillayute River station, know the main geographical features of the area, and greet people coming into the station. In addition to these tasks, the watchstander must keep a careful and detailed written account of what is taking place. The paperwork alone can bog a person down. The watchstander also keeps track of the weather and updates public weather information on telephone recording devices. This is, in short, a position that is going to be stressful and very busy with only one person on watch. Most of those coming to the unit straight out of boot camp have no idea how to operate radios and other items associated with the communications watch. Many have had no experience in the marine environment. The new people—or "boots," to use the nautical term—must know the communications routine before they can move on to any other qualification, or even go on liberty. What is amazing is that in most cases they learn all of the above within two weeks! Even more amazing is that they do it so well.

One of the important pieces of paperwork in the communication room is a worksheet the watchstander must maintain. When a distress call is received, the watchstander begins a litany of questions, reading from the worksheet, beginning with "what is your position?" followed by "number of people aboard and put on your lifejackets," and then "what is the nature of your distress?" Then the list goes on to cover other details. This worksheet gives the Quillayute River station as much information as possible quickly and in a standard manner. This standardization makes it easier for other units that may have to assist to receive the same information. Most importantly, when, as one watchstander put it, "a bell ringer goes off" and the adrenaline surges, with the officer of the day, officer-in-charge, executive petty officer, and the boat crews all wanting to know what the hell's going on, while the radio intercom from the group wants a status report, the telephone is ringing, and the person on the radio is still yelling, the worksheet ensures that the watchstander will not forget an important fact.

The new person, called a break-in, is put on with a qualified watchstander every day and learns on the job, augmented by study from manuals. To train for those periods of stress, many training petty officers will place the break-in watchstander in one room with a portable radio and then go into another with another radio and see how the new sailor performs. At one unit, for example, the training petty officer gave the new seaman her radio and said, "Got everything?" She stated she had and went into the next room. The instructor shook his head. He picked up the radio and said, "This is a drill. Uh, Coast Guard I've got a problem here. I need some help." A long silence. Finally, the seaman came into the room looking a little sheepish.

"Forget something?"

A nod from the seaman. She had forgotten the work sheet.

"How can you say you're ready if you don't have that sheet nearby? Now, are you ready?"

Over the next ten minutes the instructor proved obstinate, kept quiet at times when the new seaman tried to establish contact, made his voice difficult to hear, and at times yelled over the radio. Then he broke off the drill and told her to come back into the room.

The trainer asked if she recognized what she was doing wrong. Then he pointed out other items she'd missed. Finally: "I wasn't picking on you. Everything I did during the drill I have heard on the radio during cases." The training officer then told the seaman to practice some more. Listening to some tapes made during cases, you can learn the truth of the petty officer's comments. In one tape, a hysterical woman in distress is screaming curse words that some sailors do not use. Between the hysterical outbursts, the calm voice of the watchstander can be heard trying to get information.

Communications watchstanding is not considered one of the best duties aboard a station. Once the new break-in is ready, the sailor must face a board consisting of senior petty officers. Everyone must pass the board. They remain at the watch until they do and are not allowed to leave the station until they successfully pass the board. The results of the training are entered into the person's training record book.

A note on the training record book: The book gives the dates and initials of the person who signed off for a certain part of the required training. This, at first glance, may seem like a good idea. One petty officer mentioned that it was good to be able to see a person's training book and to recognize where the person needs work, but "the way the Coast Guard seems to operate, the first thing the officers do after a major accident is grab the training book to see if anyone screwed up in the training. It is another way they try to hang us." The petty officer's words would be prophetic.

While the communications watch is not a desired one, its very nature makes it one of the more important duties at the Quillayute River station. It is difficult for one who has not been in the position to know the amount of anxiety that floods into a person at the radio when the call "Mayday" comes blasting out of the radio's speaker. That simple word is seldom, if ever, said in a quiet controlled voice. Mayday is an internationally known last desperate call for help. When Mayday comes over the radio at a station everything seems to stop for a second; then the wail of the SAR alarm is heard and the thud of boots running to the boats, and people running to the communications room to find out what is happening.[1] An overheard warning from one salty watchstander to a newly qualified communications watchstander sums it all up: "Don't be a hero. If you have questions, pipe [page] someone. If you screw up, you can kill someone."

Once past the communications watchstander hurdle, the new arrival at the Quillayute River station can begin working in the areas of either deck or engineering work. If the person is at pay grade E-2 or E-3, he or she will again stand communications watch, but on a regular duty rotation rather than every day.

The next important qualification after communications is boat crew. The new person is fitted out with survival gear that is stowed in a large bag, called either an SAR or a ready bag. Each duty day the bag is kept near the ready boat where it is easy to grab when running to the boat on an SAR alarm. As with the communications training, the boat-crew qualification has a booklet that explains the duties and a boat-crew qualification guide to be checked off and dated by a petty officer.

Again the training is a combination of on-the-job and studying manuals. For the weekend boater, boat crew member may not seem like such a difficult position, but working a 44-foot motor lifeboat is a far cry from a pleasure craft. A warning in the motor lifeboat operator's handbook best illustrates the differences. Located directly behind the coxswain flat is a "coxswain guard screen of vinyl coated chain link fence material" that is for protection. Shortly after the handbook description of this screen is a very large warning that states: "This screen will not withstand the force of a parting towline which can generate as much as 450,000 foot pounds of force. (By comparison, a 30.06 rifle generates only 3,000 foot pounds of energy.) It is only effective against loose gear flying free through the air. Maintain your catenary, and WATCH THE LINE."[2]

Most of the work of becoming a coxswain and a surfman must be accomplished through years of experience on a boat. This means hours and hours of drill in towing boats and rescuing people in the water, plus fighting fires from a motor lifeboat. It also includes the actual working of cases involving towing, plucking people from the water, and fighting fires.

For those who have never ridden on a motor lifeboat, the boats tend to roll and pitch a great deal, even in moderate seas. Few people who have never been to sea and are put on a motor lifeboat escape seasickness. When the seas are huge in a bad case, even old salts can suffer from mal de mer. Crew members of a small-boat station look forward to how "boots" react to their first time under way, especially in the seas of the North Pacific Ocean. The seas and boat have other effects on the human body. The pounding and slamming cause a great deal of stress on the knees and back. If you lock your legs in a motor lifeboat that is taking heavy seas, you can seriously damage your knees. Spend a few hours in five to eight feet of Pacific swells and your body will ache.

One of the unique features of boat operations in the Thirteenth Coast Guard District is consistently high surf. Most commanding officers or officers-in-charge of the small-boat stations in the district try to get their crews drilling in the surf when conditions permit. After all, it is only through working in this extreme maritime environment that crews can begin to understand and prepare themselves for the night when the wind is howling and the surf is running high. Headquarters will permit stations to drill only in surf that is no higher than 15 feet. Even this limitation cannot prevent accidents from happening due to the nature of working in the surf. All it takes is the surfman or coxswain to misjudge something, or a wave to do something out of the ordinary, for the power of the surf to exert itself. Prior to 11 February 1997, for example, Master Chief LaForge had a window blow out of the CG 44363 during drills. Remember, the windows of the 44-footer are one-half-inch thick. Instead of shattering when it blew, as is usual, the window came out in one large chunk, which could have caused serious harm had it hit LaForge.[3]

During the winter months the huge surf never seems to go down. Looking at the surf crashing on the beach can be daunting. MK2 Tom Byrd, after he retired from the U.S. Coast Guard, said, "I didn't mind getting underway in the summer, but in the winter I dreaded it. I used to sit in the comms [communications] room . . . and watch the waves break over the bar and just hopin' we wouldn't get a case. The waves were the worst I've seen anywhere." If you can cut through the macho facade of those who work on motor lifeboats in the Pacific Northwest, most lifeboat people will tell you the really large waves, especially at night, can be daunting. During very black nights you cannot see the big waves until they are almost on you, but sometimes you can hear them: "They sound like a freight train approaching you." A Quillayute River coxswain said that one night he "met God at Wash Rock."

At the Quillayute River station, all crew members learn enough first aid to be first responders. Those who receive additional training are the station's emergency medical technicians (EMTs). On 11 and 12 February 1997 the station EMTs were DC2 Michael W. Keller, MK3 Matthew E. Schlimme, and SN John A. Stoudenmire III. BM2 David A. Bosley had been qualified, but had let his qualifications lapse. At Quillayute River the crew members must also know about firefighting.

Drilling and becoming proficient with the motor lifeboats, first aid, and firefighting can take up most of a crew's hours. At the small-boat stations, however, the crew must also practice a myriad of other duties. A petty officer at another station said, "I came into the Coast Guard to drive boats and save people, but I am not getting enough time under way. Instead we are painting and drilling on other things." In addition, all the maintenance of the boats and station must be accomplished by the crew. Spend time visiting stations in the Thirteenth District and you will see crews come in off the motor lifeboats exhausted, their faces showing the effect of hours of sun and wind, and before they can secure, they will refuel and then scrub down the boat with soap and fresh water. All too often these tired crew members then have to change out of their exposure suits and go back to working on a boat that needs painting or mechanical work.

One of the pressing problems of a commanding officer, or officer-in-charge, is how to get enough hours in the day to do all the training required to keep a crew qualified in the myriad tasks required; get the necessary maintenance work accomplished; be prepared for inspections; and finish the growing amount of paperwork. One of the consequences of all these tasks is the number of hours a small-boat station crew must work a week. In most cases, a crew must work at least 89–90 hours a week, with periods of 100 hours a week! Most crew members say that the first day off of their two days of liberty is spent in catching up on sleep. One woman at a West Coast station said, “The district says they want to stress family. I am so tired when I come off duty that I must sleep most of my first day. Then they go and recall us for painting and getting ready for an admiral’s inspection.”

7 U.S. Coast Guard Group/Air Station

Port Angeles, Washington

While I learned about the people and equipment of the Quillayute River station, approximately 70 road miles to the east, at Port Angeles, the men and women of the Group and Air Station began their day. When a Group is collocated with an air station, the commanding officer of the air station is in charge of both the Group and the air station. The Group is considered the senior command. The immediate superior command of the Quillayute River station is Group Port Angeles. In 1997, Captain Philip C. Volk commanded Group and Air Station Port Angeles.

A native of Beaver Falls, Pennsylvania, and a 1971 graduate of the U.S. Coast Guard Academy, New London, Connecticut, Philip C. Volk remained at the academy after graduation for about two months to teach sailing. He then reported to Newport, Rhode Island, for the precommissioning detail of the cutter *Munro,* a 378-foot high-endurance cutter. Later, he proceeded to New Orleans for the commissioning, thus becoming a "plank owner" of that cutter. He spent almost two years as a deck watch officer. From there Phil went to flight school at Naval Air Station, Pensacola, Florida, and about 11 months later earned his wings.

Volk's first duty as an aviator was flying helicopters at Group and Air Station Cape May, New Jersey. After about three-and-a-half-years, his next assignment took him to Mobile, Alabama, to what was then called Ship-Helicopter Division, where he "flew off icebreakers in the Antarctic and Greenland for just shy of two years." Phil Volk came to Port Angeles Group and Air Station for the first time in 1979 as a duty pilot and remained for two years before being sent to the academy as the senior company officer on the staff of the Commandant of Cadets.

Phil's assignment at the academy lasted for four years. While stationed in New London, Volk went to night school and received a master's degree in management from Rensselaer Polytechnical Institute.

In 1985, Phil again returned to Port Angeles, this time as operations officer. While serving in the Pacific Northwest, Volk was selected for command and left Port Angeles to command Air Station Detroit, Michigan. He was then selected for the Air War College at Montgomery, Alabama.

Upon graduation from Air War College in 1991, Volk received his first assignment at U.S. Coast Guard Headquarters, Washington, D.C. He spent his first year in headquarters as deputy chief, Congressional and Governmental Affairs Staff. The next three years of service at headquarters was as assistant chief, and then as chief of officer personnel for the U.S. Coast Guard.

Captain Volk then returned for an unusual third tour of duty at Port Angles in 1995 as the commanding officer of the Group and Air Station. Captain Volk is noted for how much he cares for the people who serve under him. Lieutenant Commander Ed Kaetzel, a pilot at the Port Angeles Air Station, said, "He really cares about his people. He really cares about the Coast Guard. It's hard to think of something that does him justice. He's the finest commander I have ever worked for."[1] In the early morning hours of Wednesday, 12 February 1997, Captain Volk had 26 years of active service, with one previous Air Station command, and 14 years of experience in flying rescue missions in helicopters.

The command and control structure of the Group and air station in February 1997 consisted of Captain Volk; Commander Paul Langlois, executive officer (XO); and Commander Raymond J. "Ray" Miller, the operations officer (OPS).

Commander Paul Langlois, a 1976 graduate of the U.S. Coast Guard Academy, first served aboard the cutter *Ironwood* in Alaska. In 1979, he graduated from flight school, and his first aviation assignment was at Air Station San Francisco. Subsequent tours of duty took him to Mobile, Alabama; Los Angeles, California; Washington, D.C.; and, in 1994, to Port Angeles.[2] Commander Langlois is noted for his love of bicycle racing; he enters races whenever possible. The endurance he gathered from this sport would be in great demand during the early morning hours of 12 February 1997.

In 1997 Commander Raymond J. Miller served as the operations officer of the Group. Commander Miller graduated from U.S. Coast Guard Officer Candidate School, Yorktown, Virginia, in 1978. His first duty station as an officer was as a duty watch officer in the National Response Center, Washington, D.C. He graduated from flight school in 1981. His first duty station as a pilot was at San Francisco from 1981 to 1985. Other assignments were at U.S. Coast Guard Headquarters and Miami, Florida. He reported to Port Angeles in 1994.[3]

In 1997, the U.S. Coast Guard Air Station, Port Angeles, had 16 officers and 59 enlisted personnel, while the Group had 4 officers and 50 enlisted personnel. Many of the officers of the air station, however, have Group management responsibilities. Lieutenant David C. Billburg, a pilot assigned to the air station, for example, is also the communications officer, public affairs officer, training officer, and classified material custodian for the Group.[4]

The air station had three HH-65A *Dolphin* helicopters assigned to it. The *Dolphin* entered service in the U.S. Coast Guard in 1985, the first of the new machines going to the service's New Orleans air station.

Some controversy arose in the selection of the Aerospatiale Helicopter Corporation, a French firm, although the helicopters were manufactured in the company's Grand Prairie, Texas, plant. Much of the debate centered on a mechanically plagued engine, a small payload, and a machine that was outside of the Department of Defense (DOD) inventory so that the U.S. Coast Guard could not obtain parts from other military services. A writer in 1991 noted, "In all its 75 years of aviation from

1916, the U.S. Coast Guard has relied almost wholly for its aircraft upon those types already selected for service with either the U.S. Navy, or in odd cases the U.S. Air Force. The selection of the HH-65A . . . proved a major break with the tradition."[5] Later, Admiral Paul Yost, commandant of the U.S. Coast Guard from 1986 to 1990, would say, "I have made a rule of thumb as the commandant that I will never again buy a helicopter or an airplane that was not a DOD supported piece of equipment."[6] By 1997, most of the problems plaguing the HH-65A had been worked out.

The contract for the service's short-range helicopter required a machine that would take off, cruise outbound at 1,000 feet in excess of 100 knots, travel 150 nautical miles from base, hover for 30 minutes, and then pick up three 170-pound survivors, return to base, and have remaining 10 percent of the available fuel, or 20 minutes of fuel, whichever is greater, as reserve fuel at shutdown.[7] The cost of the helicopter was $3,067,000.

The HH-65A has two Textron Lycoming engines and Rockwell Collins avionics.[8] The helicopter has a "computerized flight management system that integrates communication, navigation, and flight control systems. . . . Precise search patterns may be flown automatically, freeing the crew to concentrate on sighting search targets. At pilot direction the system will bring the aircraft into a stable 50-foot hover facing a selected object."[9] A pilot can override everything and fly the helicopter without the computer.

The helicopter has a maximum gross weight of 9,200 pounds and an empty weight of 6,092 pounds. It is 38 feet, 2 inches in length. The *Dolphin* has a maximum speed of 165 knots and a cruise speed of 125 knots. The maximum range of the helicopter is 400 nautical miles and the short-range radius of action is 150 nautical miles. The normal endurance for the helicopter is three hours. The aircraft has a cargo hook load of 2,000 pounds.[10]

The tail rotor, called a feneston, is shrouded for safety. For those of the Vietnam generation who can immediately identify the whump-whump sound of a Huey helicopter, the sound of a *Dolphin* has an unmistakable high-pitched whine that is easily recognizable.

The normal flight crew consists of the pilot in command (senior pilot in charge), who normally sits in the right seat. The right seat position is the best location for a pilot to see the rescue hoist. The safety pilot (copilot) sits in the left seat. An enlisted aviation flight mechanic sits in a seat behind the two flying officers. The seat moves from the left to right of the helicopter on tracks in the floor. When the HH-65A came into service the machine carried a crew of three. Soon thereafter a serious problem developed from an operational deficiency in the HH-65A and most other helicopters. Prior to the acceptance of the *Dolphin,* the standard helicopter in the U.S. Coast Guard was the amphibious HH-52A, one of the few helicopters that could land in the water. The HH-65A may do many things, but it cannot put down upon the sea. To understand the solution to this problem, it is necessary to briefly examine a shipwreck that happened in 1983.

On a Thursday evening, 10 February 1983, the *Marine Electric* sailed from Norfolk, Virginia, for Brayton Point, Massachusetts, into threatening winter weather with 25,000 tons of pulverized coal and a crew of 34. By morning of 11 February, off Virginia's eastern shore, the seas were between 20 and 40 feet, with winds howling at 60 knots. The *Marine Electric,* straining against the weather, took green water over the decks. At midnight the skipper of the ship had the holds checked to ensure the cargo was secure. Back came the harrowing report that water was flooding into the holds.

At 4:00 in the morning, on Saturday, 12 February, the *Marine Electric* sent out a call for help that was received by the U.S. Coast Guard. A helicopter from U.S. Coast Guard Air Station, Elizabeth City, North Carolina, departed for the scene. By the time the helicopter, an HH-3F, arrived, the ship had sunk and the crew was desperately fighting for their lives in the frigid Atlantic.

The U.S. Coast Guard helicopter lowered the rescue basket, but the sailors in the water were too weak from hypothermia to grab the basket. The pilot of the helicopter, Lieutenant Scott Olin, recognized that he could not perform a rescue with the equipment available and immediately requested assistance from U.S. Naval Air Station Oceana for a navy helicopter with a rescue swimmer.

The navy did not have a ready crew on the weekends and had to recall a pilot and crew. The navy helicopter joined the U.S. Coast Guard on scene at 6:05 in the morning, with U.S. Navy rescue swimmer Petty Officer James McCann aboard the aircraft. Petty Officer McCann entered the frigid 40-foot seas and swam until exhausted. The weather was so bad McCann's face mask froze. Despite his Herculean efforts, he could only rescue three people; 31 crewmen of the ship perished. McCann was awarded the Navy and Marine Corps Medal for his heroic efforts.[11]

The Congressional House Merchant Marine and Fisheries Committee convened hearings on why the U.S. Coast Guard was unable to assist people in the water. Congress then mandated in the Coast Guard Authorization Act of 1984 that the "Commandant of the Coast Guard shall . . . establish a helicopter rescue swimmer program for the purpose of training Coast Guard personnel in rescue swimming skills."[12]

Rescue swimmers are trained to *jump,* or be lowered, from helicopters and swim to a person in trouble and then assist in their being hoisted aboard the helicopter, or stay with the person until they can be rescued. The rating chosen for this duty: aviation survivalman (ASM).[13] Not surprisingly, people in this field tend to be extremely fit individuals. They must be able to swim in roiling seas and then work with people who may very well be in panic and fighting them.

The school for this rating was originally with the U.S. Navy, but in 1998 the program split from the navy and is now at the U.S. Coast Guard Aviation Technical Training Center, Elizabeth City, North Carolina. The course of instruction is 16.5 weeks. To attend the school, the person must pass a flight physical and complete a four-month airman syllabus at a U.S. Coast Guard Air Station. In addition, the candidate must pass a physical fitness requirement of 42 push-ups in 2 minutes; 50 sit-ups in 2 minutes; 5 pull-ups; run 1.5 miles in 12 minutes; and swim 500 yards in 12 minutes.

In addition to passing classroom work, to graduate from the school the person must be able to do 50 push-ups in 2 minutes; 60 sit-ups in 2 minutes; 5 pull-ups; 5 chin-ups; run 1.5 miles in 12 minutes; swim 500 yards in 12 minutes; swim a 1,500-yard gear swim; and swim an 800-yard buddy tow in 20 minutes. The person must also attend a three-week EMT school and be able to pass all the academic requirements to become an aviation life-support technician.

While in a helicopter, the ASM will normally be dressed in a wet or dry suit carrying a facemask, fins, and snorkel. The swimmer also has a Tri-SAR harness, with signal flares, radio, knife, flashlight, and strobe light. Lastly, the ASM has an EMT kit.

By 1997, aviation survivalmen routinely rode in the *Dolphin.* The seat for the ASM is to the right rear (facing forward) of the cabin and is little more than a cushion on the floor, with a safety belt. The decision to deploy a rescue swimmer is the aircraft commander's, but "ultimately it's the swimmers choice." Generally, the crew discusses options and comes to a consensus.[14]

The color scheme of the *Dolphin* is Coast Guard red, with black numbers. The HH-65A is routinely deployed to cutters, and the color scheme can help to quickly locate the helicopter when it is used aboard icebreakers in the Arctic and Antarctic. The call numbers of the *Dolphin* begin with 65. (The call numbers of the larger and longer range HH-60 *Jayhawk* begin with 60.)

The U.S. Coast Guard pioneered the use of helicopters in maritime search and rescue.[15] At the heart of using a helicopter to rescue a person in the water if the machine cannot land in the water is the use of the rescue hoist. The description of the hoist operation is simplicity itself. The pilot in command, or the pilot assisted by the computer, brings the HH-65A into a hover near the person to be rescued. The aviation flight mechanic moves his chair to the far left. The interior of the helicopter is so small that most people must move in a crouch within the machine. The mechanic attaches a safety line—sometimes called a gunner's belt—to himself, slides back the door, and locks it open. To gain some extra headroom, the top of the helicopter near the hoist can be opened and locked back, thus allowing the mechanic to stand up. Near the door is the hoist control. The mechanic then guides the pilot over the helicopter's intercom to the best location with such commands as, "move left 10, back 5," and so forth. Once in position, the mechanic can use what is known as a trail line to help guide a sling-like device to the person, or a rescue basket, or a rescue litter, on which a person can sit or lie down. The litter is stored aft of where the rescue swimmer sits. The controlling of the hoist is accomplished by the mechanic; once the person is to the door, the rescue swimmer assists. All crewmen have first-aid training and begin helping the people rescued. When using the trail-line method, a long trail line is lowered to the person, who then helps guide the basket down. The rescue hoist has a 600-pound lift capacity and has a total useable cable length of 245 feet. The number of people that can be hoisted aboard the helicopter depends on the fuel load of the aircraft, but a realistic number is three.

All of this is straightforward and relatively simple for a highly skilled and trained crew. As the helicopter approaches a boat with a person to be hoisted, the pilot in command will transmit a checkoff list to the boat. Once into the hover, the flight mechanic guides the pilot into the best position and then begins to lower and raise the hoist.

When the winds are howling and the visibility low, or in total darkness, and the seas are high, it is another story. The winds can whip the rescue basket aft and the pilot and mechanic must watch aft so that the cable and basket do not foul the rear rotor area. Sitting in a *Dolphin* you are struck by the limited vision aft. The mechanic and pilot must judge the velocity of the wind and time the rescue device with the seas. The coordination between pilot and mechanic is extremely important and can be severely tested. It is for these difficult periods that crews practice, practice, practice for the bad weather hoists.

At 2:00 P.M., 11 February 1997, the new duty section at Air Station Port Angeles reported for their briefing. Aviation Survivalman (ASM1) Charles S. "Chuck" Carter recalls that the first surprise he received that day was seeing two commanders, Langlois and Miller, doing the briefing. It is very rare that the executive officer stands duty, while the operations officer only occasionally stands a duty rotation. On this day, however, because most of the pilots had to attend a training session on total quality management, Commanders Langlois and Miller agreed to take the duty. Thus, another of the "what ifs" comes into the story. Both Langlois and Miller were very experienced pilots and—critical to the outcome of the story—Ray Miller is considered an expert in U.S. Coast Guard aviation with the use of night-vision goggles. If the two officers had not been on duty, the outcome of events might well have been entirely different.

Chuck Carter recalls that at the briefing he learned there would be a night training flight that would have the rescue swimmer aboard. He went about his normal duties until the training flight.

ASM1 Carter in 1997 had over 10 years of active duty in the U.S. Coast Guard. Originally from Maine, Chuck said, "I can't really tell you why I came into the Coast Guard." His first duty station after boot camp was at a small-boat station in Boston. "I wanted to be a boatswain's mate, but found out boatswain's mates scrape and paint boats and then when they are finished, they do it all again.

"When I was younger, I liked working out. I liked it a lot. Someone told me if I went ASM they paid you to work out. I said, '*Really?*'

"I was told I would have to spend nine weeks swimming if I was to be a rescue swimmer and I could not even stand to put my face in the water. I made it. I worked hard. I prayed hard. By the grace of God and a lot of hard work, I made it."

Chuck Carter is the picture of a fit ASM. After graduating from ASM school, Chuck's duty stations included air stations at Cape Cod, Massachusetts; Sacramento, California; Kodiak, Alaska; and now Port Angeles. When asked whether he liked duty in Port Angles, he replied, "I love Port Angeles. This is good. I had a friend who told me he would be a 20-year seaman if he could be stationed on Lake Tahoe. I would be willing to stay here as a first class and the life I'm living. I got five kids. I got a garden in. Life is good. I stand one [day] in four [days] duty."

At 5:50 P.M., prior to Chuck Carter's night training flight, Telecommunications Technician Third Class (TC3) Gina Marshall reported for duty to the Group's communications center. She would not be relieved until 5:30 A.M.; her normal watch period stretched for 12 hours. Gina enlisted right out of high school and is from Boise, Idaho. She came into the U.S. Coast Guard because "it was not in the Department of Defense and I thought it was better to save people than bomb them." After finishing boot camp, she served on the cutter *Mellon* for two and one-half years before attending telecommunications school. She arrived in Port Angeles in May 1995, as a TC3 right out of school. She likes duty in Port Angeles and met her husband in the town.

Shortly after TC3 Marshall came on duty, at 6:00 P.M., Commander Langlois, Commander Miller, an aviation flight mechanic, and ASM1 Chuck Carter climbed aboard HH-65A, number 6589, for a night boat operations training flight down the Strait of Juan de Fuca. This training would entail practice with approaching, lowering and raising the rescue basket, and possibly putting Chuck Carter in the water and retrieving him.

Commander Miller had not flown for at least 10 days, as he had been on temporary duty and leave. Furthermore, the 6589 had just returned to Port Angeles from overhaul at the U.S. Coast Guard's aircraft overhaul facility at Elizabeth City, North Carolina. One of the new items on the 6589 was lighting compatible with night-vision goggles. It was the only helicopter at the air station so equipped. (The instrument panel lights were muted to make it easier to use the night-vision goggles.)

ASM1 Chuck Carter remembers they "set up weather parameters. If the ceiling got below 500 feet we were not going to go. We started out flying and the ceiling got below 500 feet, so Commander Miller called the boat and told them we were not coming."

One of the largest changes I have noticed in the U.S. Coast Guard aviation program is how the concept of team coordination is instilled in pilots and aircrews. Barrett T. "Tom" Beard, a former U.S. Navy and U.S. Coast Guard pilot with 6,000 hours in military fixed-wing aircraft and helicopters, and the author of a book on the history of helicopters in the U.S. Coast Guard, said, "In the 1970s, the senior pilot was considered in charge and you shut up and did what he said. You had no input at all." Today, pilots brief their crews as to what is going to take place and what they expect to do. They actively support the crew's input to whatever is going to happen. This procedure makes for a much safer environment, especially when there is apt to be some dangerous work.

The helicopter returned to Port Angeles, and on the way back "they said, 'Chuck, we're going to drop you off and do some night-vision training.'" The helicopter landed at the air station. Carter hopped out of the 6589 and shed his flying gear, while the other three went out to practice for at least an hour with the night-vision goggles.

Yet another of the "what-ifs" comes into the story. Commander Langlois had very little experience in the use of night-vision goggles. Commander Miller, on the other hand, is considered an expert in their use in helicopters. Much later, Ray Miller would state, "The practice with the night-vision goggles would allow Commander Langlois the confidence to use them during the rescue." If the scheduled training flight had not been scrubbed and Langlois had not had the time to use the night-vision goggles, would the rescue attempt have turned out the way it did? After about an hour, the 6589 returned to the station and the crew secured the helicopter.

ASM1 Chuck Carter ate his dinner, then did some odd jobs and turned in for the night. Carter found the sleeping area being remodeled and was put in a room with the JOOD (junior officer of the day). "I started talking to him about God and we talked back and forth. We ended up staying up late. It was almost midnight before we went to sleep."

After his return from the night training flight, Commander Langlois had a late dinner and then did what most executive officers must do: paperwork. He finally retired about 10:30 P.M. and fell asleep about 11:30 P.M.

8 Evening

At any small-boat station that has pride, the crew is extremely anxious for visitors to see their boats. BM3 Paul Lassila took me into the boathouse and guided me through all of the compartments of one of the station's motor lifeboats. In the engine room I saw gleaming brass and could run my finger around fittings without finding dirt.

Paul and I sat in the forward compartment discussing his background and mine and his views of the "new" Guard as compared to my observations of the old Guard. Eventually it was time for the noon meal and I faced my largest hazard on this project: the meals are designed for young people with an appetite, not someone who is shaped like a pear.

While Paul and I talked, Fireman Apprentice (FA) Zandra L. Ballard, in the starboard duty section, who would come off duty at 1:00 P.M., spent the morning doing "some of my engineering PMS [preventive maintenance schedule]. One of those jobs was to PMS the gym floor by putting wax on it. Little did I know that a little goes a long way. The floor was so slick that you just kept sliding all over the place. The guys were all teasing me because they were going to be playing the locals in a basketball game that night and asking if this was my way of making sure the Quillayute River team won."

Zandra Ballard, known at the station as "Z," from Maine, tried to enter the U.S. Coast Guard right out of high school, but was under the minimum weight requirements. She finally enlisted, at the age of 23, on 6

May 1996. She joined because "I had grown up with it around me my whole life. I knew that each and every person makes a difference and that you would be known for being a person, not just another number. I like the whole mission of the Coast Guard and the saving lives."

Later, she would say, "When they told me in boot camp that I was going to be stationed at motor lifeboat station Quillayute River, in La-Push, Washington—they could not even pronounce the name of the station—I was almost horrified." Zandra would recall, as have other sailors for a long time, "I had asked for the East Coast; I got the West Coast. I had asked for a cutter; I got a station. It did not look like things were going like I had wanted them to at first. Then I asked to be a seaman and—voilà—I was not. I was put as a fireman. I was engineer illiterate at the time. So, here I was wondering what was going on and just saying in the back of my head, okay, here goes a whole new adventure. I was thinking the Coast Guard sees some reason to be doing this to me, so I have to accept it."

When Zandra arrived at the Port Angeles airport, "There were several of the crew—Miniken was one of them—to greet me. On the way back to the station, I instantly started bonding with all of them. I remember getting to the station and saw the Master Chief outside pulling weeds. He came up to me and said 'hi.' I knew right then and there that being here was going to be a great time."

The petite Zandra would also recall her first time on a 44-foot motor lifeboat. "I had grown up on the water and had never gotten seasick. Okay, wake-up call. I think that I have turned every shade of green possible. I remember Schlimme gave me some crackers to help the sick feeling." Master Chief LaForge, however, would later say Zandra had one of the strongest stomachs of the crew.

Seaman Apprentice (SA) Benjamin F. Wingo, 19, from Bremerton, Washington, who was assigned to the ready boat crew as a break-in, entered the U.S. Coast Guard in 1996. He had wanted to join the U.S. Air Force to work in their search and rescue organization, but could not be guaranteed the field. He chose, instead, the U.S. Coast Guard. Ben would later say he never remembered "ever worrying much about anything." The 6-foot-4-inch, 220-pound blonde played outstanding high school soccer.

Ben earned partial college scholarships, but could not afford the difference. So, as many others before him have done, Ben decided to enter the military and obtain the college fund most of the services offer for those who enlist. His first duty station out of boot camp was Quillayute River. He later said, "I wanted to drive small boats and rescue people. That would have been awesome."

Ben considers himself lucky in life. He also admits that it is hard for him to take anything seriously. In other words, he has a devil-may-care attitude. Ben related that one time when he was fooling around, one of the people at Quillayute River thought he had trouble paying attention. Some of the crew mistakenly took this easy attitude for slow thinking.

Despite the isolation, Ben liked his first duty station. He liked the work outside, maintaining the boats and learning how to become a boat crewman. Ben also liked playing sports and partying in Forks with friends. Interestingly, the previous weekend, Ben and other crewmen during their constant training underwent instruction in pyrotechnics, including how to work the red pencil flares carried in the pyro vests all crewmen wear while under way.

After lunch, Bosley asked BM2 W. Brent Cookingham, of Tacoma, Washington, to fit me out with the necessary survival gear in anticipation of riding a boat during the station's constant boat drills. Brent, one of the station's three surfmen, is described by a crew member as "an old seadog." Like sailors of my time, his arms show the results of the tattoo artist. Brent told me that he and his brother ran a salvage tug in California before he came into the U.S. Coast Guard. "If you've seen the movie *Splash,* the scene where there is a diver aboard a tug, that was our tug."

Brent gave me all the necessary paraphernalia to survive in the cold waters of the North Pacific: an antiexposure coverall, warm thermal underwear, socks, rubber boots, cap, gloves, and a helmet to protect my head. I received a SAR bag to stow all the gear.

Petty Officer Cookingham and I talked about how the coast guard now spends so much time on environmental practices. While a good policy, the added paperwork places additional work upon already overburdened senior petty officers. Cookingham picked up a thick manual. "We're required to post these anywhere there are hazardous materials. Even the cleaning locker in the barracks has to have one of these so everyone knows what is in the cleaning material." Today, Cookingham informed me, the nonrates think nothing of donning a respirator when opening a can of paint. This is a far cry from the old Guard, when you pried open a can of paint, stirred it, and then went to work.

I went to the communications room and talked to FA John D. DeMello, 18, right out of boot camp and at the unit less than three months. John is from Hawaii, "the Big Island," and the town of Kealakekua. John told me he came into the U.S. Coast Guard "on a bet between a friend. We didn't want to just stay on the island and do nothing like a lot of the kids do, so one day we said, 'You want to join the military?' He said, 'I will if you will.' I said, 'I will if you will.' Nine months later I was in boot camp.

"My dad actually knew about the Coast Guard. I was leaning toward the army, but he was in the army. He said the Coast Guard was on a lot of the islands. They did a whole bunch of things. I didn't know they were a military service."

"Washington [State] was my last choice and somehow I got it. I wanted Hawaii, obviously. I had never been to the mainland. Then I put in for Northern California and Oregon. For some reason I did not want Washington, but I don't remember why. It ended up being one of the best experiences of my life."

I said that Quillayute River certainly must be a big change from his home in Hawaii. John did say the Quillayute River area reminded him of "Hilo, the other side of the Big Island. It is rain foresty, like Hilo. The rain, small towns, kinda reminds me of home."

John mentioned he was amazed by the number of birds in the area. I noticed an eagle perched in a nearby tree and took the spotting telescope in the communications room to watch it for a few minutes.

At approximately 4:40 P.M. a National Weather Service coastal weather forecast message was issued and sent to the station. The message stated that a gale warning was in effect from Cape Flattery to the mouth of the Columbia River and out 60 nautical miles from the coastline. Winds would be shifting to northwesterly 35 knots, gusting to 40 knots. The seas were forecast to build to 14 feet. For Wednesday, 12 February, the forecast indicated northwest winds of 30 knots and combined seas of 18 feet. When individuals view the forecast on the computer, they generally leave their initials to indicate they have read the information. Later, the investigation into the incident would show that at 5:40 P.M. someone with the initials "DAB" viewed the weather forecast.

At approximately 5:30 P.M., BM1 Placido, as the surfman on duty and senior duty officer, and Bosley, the officer of the day, went to the hill to do a last light bar check. The weather and seas did not seem bad, so Jon Placido left for his home in Forks.

"We had some wind chop. . . . As far as the swell is concerned, there was none to speak of. From the looks of the bar and what Bosley told me about the weather, there was no need for me to stay aboard." Bosley failed to inform Placido of the weather forecast.

I spent the evening talking to crew members about their lives, their likes and dislikes, and their hopes for the future. I talked to one petty officer who said he was slowly becoming disillusioned with the Coast Guard because of the number of hours he worked. As many said during my interviews at other stations, he was so tired by the time liberty came, all he wanted to do was go home and sleep. He really did not have much time for his family.

I sat on the couch on the rec deck and talked to MK2 Thomas L. "Tom" Byrd. Tom had 18 years in the U.S. Coast Guard and planned on retiring in Florida. One of his duty assignments is unique. For a one-year period, from October 1991 to October 1992, his duty station was a LORAN (Long Range Aids to Navigation) station on Iwo Jima, the volcanic island in the Pacific that was the site of one of World War II's more vicious battles. Unlike many who are assigned to a unit in a historical location, Tom became interested in the battle for the island. Along with another U.S. Coast Guardsman, he explored the many caves the Japanese used for defense. Some of the caves had not been entered since being sealed during the battle in 1944. He spoke of finding a seabag with the pristine uniform of an Imperial Marine, medicine bottles, and even some letters of a Japanese soldier. Tom also kept a very good scrapbook of his service career. He showed me an article in the Pacific edition of the service newspaper *Stars & Stripes* about his room in the station's barracks. He had created a mini-museum of the artifacts he had found and models of aircraft he had made. "I donated the models to a small museum the Japanese had on the island when I left," Tom said.

I met very few people in my visits to stations who wished to talk about history. Tom was the most enthusiastic. We spent a good deal of time talking about the history of World War II in the Pacific. Tom speaks very deliberately, and with just a touch of a Florida accent. The conversation revolved largely around various books on the subject and why there seemed to be so little interest in the Pacific theater of war as opposed to the European theater. Tom knew more about the Battle of Iwo Jima and the war in the Pacific than most college history majors. Afterward, I always looked forward to talking with Tom during visits to the Quillayute River station.

While Tom and I carried on our conversation, some of the crew members began to get ready to play basketball with some tribal members. I recall someone telling Bosley they were going to be using the gym to play with the locals. Bosley did not seem too pleased about the arrangement and said, "You know, you are responsible if they do any damage in the gym."

Zandra Ballard, even though off duty, came into the communications room and relieved John DeMello so he could play ball. "Zandra was nice like that; she always took over watches for us," John said later.

There were "three hard games" of ball; the only results I have learned is that whoever had Ben Wingo on their side won, because "he's so tall and closer to the net."

As part of the Quillayute River duty section played basketball and Zandra stood watch, Lieutenant Kenneth Schlag of the U.S. Navy, attached to the U.S.S. *Carl Vinson* in Bremerton, Washington, was sailing his sailboat, the *Gale Runner*, from San Francisco to Bremerton. Schlag had lived aboard the *Gale Runner* while in San Francisco, so in effect he was moving his home to his duty station. Also aboard the sailboat was his friend Marcia Infante. Off the Washington coast, the *Gale Runner* began to run into very heavy weather.

Later, some would question Schlag's decision to sail a boat northward along the Pacific Northwest coast in February. Maritime historian Tom Beard, a retired U.S. Coast Guard pilot, who, along with his wife, Caroline, has sailed over 160,000 miles in their 37-foot sailboat *Moonshadow*, said, "I tell anyone asking me about sailing along the Washington-Oregon coast to go either north or south from April to October. This timeframe is mostly for power boats. It is dicey to go in a sailboat northward before August. If you do try it in a sailboat from April to August, you have to be ready to quickly run into a port. From October until April, truck your boat up or down I-5." What many do not realize is that the U.S. Coast Guard, except in the Thirteenth District (Washington and Oregon), cannot prevent anyone from getting under way and going out into any type of sea, no matter how bad the weather. The service can only give the current and forecasted weather. In some cases, U.S. Coast Guard boarding officers can terminate voyages if major safety equipment is found to be lacking. The Thirteenth Coast Guard District's officers-in-charge and commanding officers can prevent recreational craft from crossing a bar if they deem the bar is unsafe. They cannot, however, prevent commercial craft from crossing the same bar. The *Gale Runner* was a seaworthy boat and Schlag departed from San Francisco; therefore there was no legal way the Coast Guard could prevent him from sailing.

Later still, some would ask why there was no legal action against Schlag for his part in the deaths. One of the reasons for this lack of action is that he broke no law in sailing northward in February. Furthermore, if boat owners knew that every time they called for help they could end up with legal proceedings against them, many would not call for assistance. Almost no recreational craft are under the command of licensed owners. Licensed mariners have different rules. If a vessel under command of a licensed owner, such as a charter fishing boat, made a poor decision resulting in death or injury, then the Coast Guard could take legal actions against that person in the form of license suspension or revocation as well as civil penalty.[1]

BM1 Dan Smock, of the U.S. Coast Guard small-boat rescue station Grays Harbor, Washington, recalls that Schlag did come into the station on 11 February 1997 to check the weather and then got under way. The bar was not closed, so there was no way to prevent the *Gale Runner* from departing. Sometime during the same night, Lieutenant Schlag, fighting the weather, decided that he must put into Quillayute River.[2]

"The crew that was out playing ball came in once and I asked how the floor was," recalled Zandra. "Miniken looked at me and said that after everyone fell on the floor a couple of times, it seemed to dry it. They all said, 'Leave it to Z.'"

Clinton "Clint" Miniken, 22, had been in the U.S. Coast Guard for eight months. A high school teacher recalled Clint as "unassuming, nice, and well-mannered." He qualified for boat crewman in three months. Clint's shipmates called him a "superb performer." One petty officer told me that he once gently chided Miniken for not doing something correctly. "It was more in the way of a joke, but he was very despondent, because he always wanted to do the best he could."

Chris Koech, who served at Quillayute River before the accident, said Miniken "was a fanatic about sports. He loved his basketball games." Chris remembered that Clint went to a community college, with a major in criminal justice, and "planned on becoming a police officer when he was eventually discharged from the Coast Guard."

Chris recalled that Clint Miniken was a "quiet person," but "did not have any trouble standing up for himself. There was a time on the mess deck when someone pressed him too much about his personal business and he let the person have it. The rest of us were laughing about it. If you were to flip him some guff about one of his teams, he would give you a friendly earful."

Koech pointed out that Clint Miniken was a relatively new member to the station when he served there, "but he was a competent crewman from what I remember of the cases we were on. He knew his job and did it without having to be prompted, and he was never a safety hazard. When we did surf drills, he was calm and professional."

Later, Zandra recalled talking to MK3 Matthew E. Schlimme. Schlimme, 24, of Whitewater, Missouri, had only 16 days remaining in the U.S. Coast Guard. Zandra said, "I mentioned to him that he should be counting down the days now until he was gone from here. He said he wouldn't believe it until it actually came to it, because anything could happen." Zandra and Matthew talked a bit about the weather. As Zandra recalls, "[Schlimme] said that the wind might be blowing, but at least it wasn't raining, so if there was a case it wouldn't be that bad."

Sometime shortly after talking to Zandra, Matthew E. Schlimme came onto the rec deck and sat down near me, and we began to talk. There are some people you meet who seem most natural with a smile on their face. Schlimme fit into this category. His nickname on the unit was "the Jackal," as his job was to make people laugh. FA Falacia Brantley would later echo what many at the Quillayute River station said, "Schlimme was a special person." BM1 Placido recalled that during breakfast he and Schlimme would look through the want ads section of the *Forks Forum,* the small local newspaper, for ads dealing with guns.

Chris Koech recalled Matt Schlimme's sense of humor. "I still laugh when I think of some of the crap that kid pulled. Talk about a dry sense of humor, and right to the bone. When Matt torpedoed you, you were sunk. He had that sense of humor. He never cut anyone to hurt them as I recall; it was all in good fun, but if you were in his sights, you were doomed. Oh, God, he was a riot."

Chris went on to say that Schlimme "had a combination of that country boy common sense, and a fine mind. I think Matt would have done very well for himself whatever he did. He was a dreamer and pretty much got along well with everyone: People liked Matt. He was smart, good natured, and a good crewman. I had no complaints with Matt; no one did I don't think. He was just a good guy."

Schlimme's sister, Angie, said Matt "was going to leave the service and move home to the family farm in Whitewater, Missouri." The Schlimmes, according to Angie, "are a close-knit family." She, her brother Andrew, and their parents live near one another. Matt, however, decided to enter the U.S. Coast Guard, wanting to move around for a while. When he was ready to leave the service, "all of us kids were going to build a house on this land," said Angie.

As with many small-town people, Matt met his future wife, Christina, at Jackson High School in Jackson, Missouri, where he attended school. He graduated in 1991. Matt married Christina in a small church in Tilsit, Missouri.

My impressions of Matt Schlimme, and those of most of the people I met, can best be summed up by his sister: "He thought more about anybody else than he did himself." His actions on 12 February would bear out this comment.[3]

Before entering the U.S. Coast Guard, Schlimme worked on tugboats on the lower Mississippi River. His first duty assignment after graduating from boot camp was aboard a small buoy tender on the Upper Mississippi River. He told me one of the things the tender did was tie up at a small town every evening. Schlimme enjoyed walking along the river and visiting the small river towns. "I really liked the small, clean towns," he said.

I told Schlimme of my forays along the Mississippi River when I worked at the Rock Island, Illinois, U.S. Army Arsenal, and he immediately hurried off the mess deck, coming back with a road atlas. We sat and looked at the names of towns along the river, seeing if we had been in the same ones. When I mentioned I needed the address of someone who had been stationed at Quillayute River, he called his wife and obtained the information for me—another example of the helpful and friendly nature of most people at the stations. This would be Schlimme's last call to his wife.

In the meanwhile, BM2 Bosley had come onto the rec deck and sat down near us. Bosley, I learned, had received orders to the cutter *Point Hobart* (WPB-82377) an 82-foot patrol boat, its home port in Oceanside, California. "These are my wife's orders," said Bosley. Seeing my quizzical look, he explained that she was from southern California and when he was last transferred he had wanted to come to the Quillayute River station. Now it was time to go to southern California.

Some time after I spoke with Bos, Master Chief George LaForge came onto the rec deck. He had returned from Port Angeles and checked the weather with Bosley. According to LaForge, Bosley did not mention the National Weather Service forecast received at 4:40 P.M.

LaForge wanted to know how I was getting on at the station. He said it looked as if there would be some good surf running tomorrow and asked if I wanted to go aboard the motor lifeboat for surf drills. At the beginning of my project, I had ridden a motor lifeboat for about three and a half hours in an eight-foot rolling Pacific swell and was sore for a week afterward. That would have been a gentle ride compared to the pounding in the surf. George must have seen my hesitation, for he said that it was okay if I did not want to go, "as safety was the most important thing." George left shortly thereafter.

Master Chief Boatswain's Mate (BMCM) George A. LaForge, the officer-in-charge of the Quillayute River station, is originally from Tacoma, Washington. He entered the U.S. Coast Guard in August 1971. His first assignment out of boot camp in October 1971 was aboard the buoy tender *White Bush* (WAGL-542), with a home port at Astoria, Oregon. Three years later, he was transferred to station Siuslaw River, Oregon. George told me he remembers pulling into Yaquina Bay, Oregon, on the *White Bush* and visiting with the people at that station. They were sitting around on the steps of the station and the guys "told me how much I would like duty at small-boat stations." He also asked his chief about duty at a station. George's chief, BMC Darrell J. Murray, said, "It would be different duty, but you would not forget it." In another twist of coincidence, or, if you will, fate, BMC Murray earlier in his career had been involved in the case of the fishing vessel *Mermaid.*

On 12 January 1961, the *Mermaid,* attempting to enter the Columbia River, sent out a distress call. The fishing vessel had lost her rudder and was drifting toward treacherous Peacock Spit. The case started out routinely enough. Then everything possible seemed to go wrong. In the end, Murray, at the time a BM1 and coxswain of a 40-footer, had the boat pitchpole, trapping Murray and another crewman beneath the upside-down hull. (The 40-foot utility boat does not right itself.) Both managed to work their way out. In the end, the service lost a 36-foot motor lifeboat, a 40-foot utility boat, and a wooden 52-foot motor lifeboat that terrible night, along with five U.S. Coast Guardsmen and two fishermen. This represents one of the largest single losses from a motor lifeboat station since the U.S. Coast Guard was formed in 1915. Thirty-six years and one month later, George A. LaForge, the man who asked Murray about duty at a small-boat station, would lose three men in a rescue attempt.[4]

In August 1977, LaForge again moved, this time to a newly built station at Destin, Florida. For all those who love lighthouses, George's first isolated duty came in 1979 at the U.S. Coast Guard Light Station, Isles of Shoals, New Hampshire. This isolated rocky island is used many times to illustrate a rocky wave-lashed lighthouse. George had two weeks of duty and one week off. To get off the island he had to row a 15-foot dinghy from the station to the boat sent out to take off the person going ashore and bring his relief. George pointed out that at times the weather was too bad and he had to remain longer at the station. At that time, isolated duty was usually for one year.

In January 1980, LaForge received orders to travel again across the United States and returned to the station at Grays Harbor, Washington.

George was promoted to chief petty officer (E-7) at this time, which meant he had to move once again. Chief LaForge received orders, in August 1981, to the U.S. Coast Guard cutter *Rush* (WHEC-723), a 378-foot high-endurance cutter, home ported at Alameda, California. LaForge received orders in July 1983 to another cutter, the 95-foot patrol boat *Cape Jellison* (WPB-95317), home ported in Seward, Alaska. He served as the executive petty officer (XPO) on this patrol boat.

LaForge returned to the Grays Harbor, Washington, station again in December 1985 as the XPO and remained there until June 1989, then once again returned to sea duty aboard the 82-foot patrol boat *Point Ledge* (WPB-82334) at Fort Bragg, California. This time, however, he served as officer-in-charge. I have heard unconfirmed stories that George, like many sailors, is superstitious when going to sea. He did not like to hear whistling on his patrol boat. This comes from the age of sail when ships were becalmed. Many times skippers would muster their crews and make them stand while the master would whistle for wind; thus whistling can cause wind and bad seas.

BMCM LaForge received orders to the National Motor Lifeboat School, Ilwaco, Washington, in March 1992. George's duty at this assignment was to be in charge of the Standardization Team. This duty required LaForge to visit all the U.S. Coast Guard stations that had motor lifeboats. He would then test each unit to see if they were in compliance with standard operating procedures for operating motor lifeboats. The STAN Team visits are not eagerly awaited by stations. Careers can be seriously hurt by the results of the testing. George probably caused more than a few boatswain's mates to dislike him because of this assignment.

In May 1996, after close to 25 years of service, BMCM George A. LaForge took over as officer-in-charge of the Quillayute River station, relieving Senior Chief Boatswain's Mate Daniel Shipman. LaForge took the assignment, which is actually billeted for a senior chief, "because no one else wanted the duty."

When I visited many of the small-boat stations one thing that struck me was the size of most of the boatswain's mates in charge. Most fit the stereotypes of big, burly men. George is a surprise. He is lucky to reach 5 feet 8 inches. Like the stereotypical master chief boatswain's mate, however, he loves to talk. George is a very good sea storyteller, especially if the sea story is convoluted and lengthy. Master Chief LaForge, like many surfmen and lifeboatmen, cares deeply about the small-boat rescue stations of the U.S. Coast Guard. Lieutenant Mike White, former commanding officer of the Cape Disappointment station, said, "George LaForge was absolutely dedicated to the professionalism of the motor lifeboat community." Further, George "was always an advocate for the troops. We often disagreed on how we would solve the ills of the U.S. Coast Guard, but George's heart was with the kids on the boats, their workload, their training, and their future." After the deaths of the three U.S. Coast Guardsmen, a surfman with many years of experience on the coast said, "We talked about who would be the worst possible person to have such a thing happen to and everyone agreed it was George. Some of us become what you might say tough, or have other ways to deal with death, but George wears his heart on his sleeve. He really cares about the people who work for him." Chief Warrant Officer (CWO2) F. Scott Clendenin, commanding officer of the Yaquina Bay, Oregon, station, said, "George has a heart bigger than he is. I don't know how it is housed in his body."

CWO2 Clendenin also remarked: "George has got good training skills." Scotty Clendenin, however, went on to say, "I'd never want to be a George LaForge. I'd never want to be stationed up at Quillayute River. George knew his number one priority was to keep up the crew's morale. Without morale you don't have training. Without training you don't have morale. George is thinking, 'I got to make sure there is something else to do here than drink; I've got to keep them interested.'"

George was not afraid to let his views be known to everyone from the Group to headquarters. I have had officers show me some of his lengthy e-mails in which he is strenuously pointing out something flawed in the system that might hurt the small-boat stations.

After a briefing by Bosley about the weather, LaForge felt the duty surfman could remain at home. Later, George would testify that Bosley did not mention the updated weather forecast. LaForge took care of some additional matters and then departed for his quarters, about a mile from the station.

By now it was time for me to turn in. I said goodnight to Bosley and Schlimme. I learned that both had the ready boat for the night and that Schlimme now had fewer than 16 days to go in the Coast Guard. He gave me one of his smiles and said, "I'm going home."

On the way to my room, I looked through the window at the weather. Wind shook trees in strong, gusty bursts. Visibility seemed low, and I could hear the surf crashing on the beach a quarter of a mile away. I dozed off wondering if I should reconsider going aboard the 44-footer in the surf. Riding a motor lifeboat is a young person's game.

Shortly after I left them, around 9:30 P.M., Bosley, Schlimme, and Fireman William C. Matthews started out to make the last evening rounds of the station before they retired for the night. Later, Matthews recalled hearing Bosley and Schlimme state "they hoped they didn't get a case that night."

BM1 Placido's standing orders were for the officers of the day to check in with him at 10:00 P.M. when he stood senior duty officer. Bosley called Jon Placido at home at 10:00 P.M. "He said he had just taken a drive around and drove down to the bar and nothing was going on. The winds had picked up a little bit. He said the weather was supposed to pick up later the next day."

After the basketball game, FA John DeMello showered and returned to the communications room to resume his watch. At 10:00 P.M., John turned over the radio guard to Group Port Angeles and trundled out the cot and his sleeping bag. Ben Wingo and Fe stopped by the communications room and talked until close to 11:00 P.M. "You probably aren't supposed to do that," said John later, "but being watchstanders, they knew it can get lonely in there and it helps to relieve the tensions."

FA Falicia A. "Fe" Brantley, 19, from Tennessee, came into the U.S. Coast Guard because she saw "a film of someone jumping out of a helicopter and rescuing people" and wanted to join an organization that saved people. When she finished boot camp she wanted to be a sailor and be assigned to a cutter. Then she learned of her orders to Quillayute River. "They couldn't even pronounce the name of the station. I wanted to cry, but I couldn't do it in front of the company commander."

It was dark when she arrived in Port Angeles and "everything was so dark: the mountains and especially around Lake Crescent. They kept saying we are almost there and then we would keep going further and further."

Eventually, Fe would look upon the "Master Chief and Miss Melva [LaForge] as my mother and father and the crew as my family." Fe had the reputation of "getting into someone's face" if she saw something wrong, or if someone said anything against the Quillayute River station. Fe was the only African American on the station and most likely the only African American in the west end of Clallam County. On 11 February, Fe was off duty and returned late at night from a doctor's appointment in Bremerton, Washington. She recalled that before going to her room for the night, she gave Schlimme a notebook someone had started of all the silly things people at the station had done. Fe roomed with Zandra Ballard.

Zandra recalls that after DeMello came back on watch, she saw Miniken as she was walking down the hall in the barracks. "He said he was going to hit the rack, as he was beat." Ballard remarked that it was too bad he had first boat. "He said that he hoped there wasn't a case, as he would die from being so tired."

Zandra then went to her room. Over two years after the night of 11–12 February, she recalled, "Me and Fe were talking for awhile and I can still hear the wind picking up around the station. Our room was on the end, so the back door kept creeping open a little and slamming shut."

. . . Part Two **Wednesday, 12 February 1997**

9 "Mayday! Mayday!"

Some time before midnight, Marcia Infante, aboard the sailboat *Gale Runner*, picked up the radio's microphone, tuned to channel 16, and called the U.S. Coast Guard Station at Quillayute River to obtain their bar report. Telecommunications Specialist (TC3) Gina M. Marshall, of Group Port Angeles, who had Quillayute River's radio guard, returned the static-filled call and had Infante switch to channel 22A. Marcia Infante switched channels and asked if there were any restrictions on crossing the Quillayute River bar. TC3 Marshall looked at her posted information in the operations room and found no restrictions. She passed the information on to the *Gale Runner*.

FA John DeMello came awake in the communications room to Infante's call of "Coast Guard, Coast Guard" and heard Marshall's response about no restrictions on the bar. The 18-year-old DeMello glanced at the station's wind anemometer gauge and watched the needle jump to 50 knots in gusts. He recalled that the last marine weather broadcast had spoken of the possibility that seas might build to 18 feet in height. DeMello, on his own initiative, called TC3 Marshall and told her the present wind conditions. DeMello also told Marshall he felt that if the Quillayute River station's officer of the day, BM2 David A. Bosley, knew of the present conditions, the bar would be restricted. TC3 Marshall told DeMello to call the *Gale Runner* and pass on the current information to

Fig. 9.1. Aerial view of entrance to Quillayute River, shot the day after the deaths, with seas still high. The U.S. Coast Guard boathouse is the largest white structure at the top of the protected area off the river. To the right of the boathouse is the station building. The cove that the motor lifeboat entered is visible on the island's left side. Photo by AN Angela Engle, U.S. Navy, U.S. Coast Guard photograph.

them. DeMello called the sailboat on the radio and asked, "Are you coming across the bar?"[1]

Infante replied, "Affirmative. We are at the entrance now." It is important to remember Infante's transmission, as critical decisions were predicated upon it. Infante did not pass on the sailboat's actual position.

DeMello radioed the current wind speed and informed Infante that the bar could be breaking, one of the most dangerous conditions facing a mariner (fig. 9.1).

On the tape recordings of the radio transmissions, Marcia Infante's voice has the tone of a young person. Part of the reason is her concern about what is taking place aboard the *Gale Runner.* In tense situations, voices tend to rise, and people speak much faster when using a radio.

DeMello picked up the station's telephone and called BM2 Bosley, who had retired for the night to the OOD's duty room (fig. 9.2). When Bosley came on the line, DeMello informed him of the *Gale Runner's* intentions of coming into Quillayute River and of the current weather conditions (fig. 9.3). Bosley told the watchstander he would be right down. After arriving in the communications room, Bosley asked DeMello if he had the position of the sailboat. DeMello said the person calling had not relayed it. BM2 Bosley ordered DeMello to call the boat and obtain a position. Bosley called the senior duty officer (SDO), BM1

Fig. 9.2. BM2 David A. Bosley in 1996 at the Quillayute River station. Photo courtesy Thomas L. Byrd.

Fig. 9.3. Approaching the entrance to the Quillayute River after leaving the U.S. Coast Guard station on a calm day. Photo by Dennis L. Noble.

Jonathan Placido, at home. He informed Placido that the station had received a radio call from the *Gale Runner* stating they wanted to cross the bar. Bosley felt the current weather conditions would make the transit hazardous for a sailboat.

"Bosley told me we had a sailboat out there who wanted to come in," recalled BM1 Placido. "I said, 'What's his problem and where's he at?' The problem we have with a lot of sailboats is that they are bucking a head sea. What they need to do is turn around.

"He said, 'Just a second.'" Bosley placed Placido on hold.

"I am thinking the boat is somewhere beyond the buoy line, but close to the whistle buoy," recalled Jon.

DeMello, "Can I get your position?"

Infante: "We're at 47–51 . . ." Static.

DeMello, "Sailing vessel, this is Quillayute River on 22, over."
A period of static-filled silence.
"MAYDAY! MAYDAY! MAYDAY! THIS IS THE SAILING VESSEL *GALE RUNNER*! U.S. COAST GUARD! WE'RE TAKING ON WATER!" Static.
TC3 Marshall: "Sailing vessel calling Mayday. Request your position."
DeMello keyed the radio intercom to Port Angeles: "Port Angeles, you got 'em?"
TC3 Marshall: "No. I'll try calling Astoria." Gina Marshall called U.S. Coast Guard Group Astoria's communications center and asked if they had heard the Mayday. Astoria had not, so Marshall then called the Tofino, British Columbia, radio and also received a negative reply. TC3 Marshall then called the Group Duty Officer.
"MAYDAY! MAYDAY! MAYDAY!"
A period of static-filled silence.
"MAYDAY!"
Another period of silence. Later, Infante would say she broadcasted about every fifteen minutes.
Marshall: "Sailing vessel calling Mayday. This is Group Port Angeles. Request your position."
Marshall then keyed the radio intercom to Quillayute River: "Request you turn up your low site. . . . Try them on your low site."
DeMello: "I already did that."
Marshall: "You getting anything?"

Aboard the *Gale Runner,* Kenneth Schlag was steering the sailboat, while Infante, in the cabin, worked the radio. Suddenly, as Infante began transmitting their position, a rogue wave struck the sailboat and knocked it down, partially carrying away the mast and damaging the rudder. When the mast went, so did the radio antenna. The wave threw Schlag into the cold, churning sea.

Schlag, tethered to the *Gale Runner* with a safety line, managed to pull himself through the heaving sea to the boat. The boat righted itself, and Schlag managed to clamber aboard. Schlag later said, "Hatches and portholes were blown out, leaving the boat flooding, the mast torn from the boat and the motor flooded out, leaving it inoperable."[2]

Marcia Infante, meanwhile, was thrown violently about the dark cabin. Marcia became aware of water rushing into the space. Groping about in the dark, she could feel the cold sea rising in the compartment. Marcia fought her way aft, struggling against the pitching and rolling boat, trying to work her way up the passageway leading to the cockpit. Blocked. She called for Schlag.

Schlag, after clambering aboard the *Gale Runner*, began to clear away the debris about the deck. Schlag left the disabled mast attached to the boat, hoping it would provide some stability. He cleared the debris blocking the entrance to the cabin and finally reached Infante. They both bailed feverishly. Marcia tried the radio again. She broadcast that they were taking on water.

Marcia continued to call every few minutes. When she looked at the GPS for the position, she saw a blank screen and thought the instrument had not survived the knockdown, so she could not give a position with the distress call. She did not know that a screen saver had caused the blank screen, so that she would only have had to hit a key to pull up the position. Marcia heard only a few garbled radio responses.

In the communications center at Port Angeles, TC3 Marshall heard the Mayday call and tried to obtain a position from the sailboat. Again, feeling the radio transmission might be from a child, her first thoughts were: "What is a school-age child doing up at this time of night?" She put out an urgent notice to mariners about the Mayday, asking for any assistance from mariners in the area.

Bosley had to process a great deal of life-threatening information very quickly on this stormy night. The radio transmission said that the sailboat was at the entrance to the bar. Next, the station received a Mayday call that the boat was taking on water, followed by a loss of communications. These facts play an important part in Bosley's action.

Bosley asked DeMello to reach to the right of the communications console, press the SAR alarm, and say over the public address system: "Ready boat crew, lay to the ready boat! Sailboat on the bar taking on water!" The announcement gives an indication of his thinking. Next, Bosley took Placido off hold on the telephone and told the surfman that the sailboat was taking on water. Bosley said, "I'm heading out the door."

Placido recalled, "I said, 'Call Master Chief. I'm right behind you.' Master Chief lives 5 minutes away, I live 15 minutes away. I'm not going to let a resource like that slide through. If I had known how bad the bar was, I would have told Bosley to wait until I could get into the station. I'm on my cell phone talking to the station pretty much the whole time I'm on the way in. There are two or three dead spots where I can't talk to the station. As I'm coming in, I'm talking to DeMello, and I keep on asking, 'Are they under way?' He keeps on telling me no. I am at Three Rivers [about seven miles from the station] and he tells me the boat is under way."

At approximately 12:30 A.M., Master Chief Boatswain's Mate George A. LaForge, asleep in his quarters, heard the ringing of the telephone. DeMello informed LaForge that a sailboat wanted to transit the bar. DeMello also told LaForge about the boat's broadcasting a Mayday and taking on water, and about the station's losing communication with the sailboat. Master Chief LaForge told DeMello he was on his way.

The SAR alarm and public address announcement woke FA Zandra A. Ballard in her room in the barracks (fig. 9.4). "I immediately jumped up, automatic instinct. Fe was getting up herself and getting on some clothes. I ran down the hall and made it to the foyer and I looked out the window and saw Bosley running back up to the station. He ran in and hollered where was his boat crew. I told him I would go and he said okay and go grab one of the guys and haul it down to the boat."

Zandra ran through the rec deck and down the barracks hall. "Just as

Fig. 9.4. FA Zandra L. Ballard. Photo courtesy Thomas L. Byrd.

I was about to knock on the door [of Miniken and Wingo] it opened and out they both stumbled. The sleep was still in Miniken's face. As they raced by me, I said to them to be safe and I would see them in a bit. I recall that Miniken's boots were not laced and he was wearing a tank and shorts."

John DeMello remembers seeing Clint Miniken and Ben Wingo running by the communications room to the boats. "Wingo stuck his hand inside the window, smiled, and kept on running."

Zandra went back to the communications room. "You could hear this woman frantically talking on the radio. The sweat was literally pouring off DeMello."

Zandra automatically started to help DeMello.

"Port Angeles called up and said to have the boat stand by as they were thinking this was a hoax. I was thinking, are they nuts? This person seemed to be in serious distress. So I got on the radio with Bosley and he said to stand by one as they were a little busy at the moment. I never got a chance to tell them to hold back. Exactly one minute later Bosley called in that they had crossed the bar safely and that it was 16 to 18 feet on the bar and 16 to 18 outside."

Zandra recalled that she heaved a "sigh of relief." She now waited to find the position of the sailboat, as "we had thought it was on our bar."

Meanwhile, DeMello had cleared out the sleeping cot and SA Trevor K. Sowder had also come into the communications room. "At 12:47 [A.M.] me and Sowder heard 'capsized and disorientated,'" Zandra recalled. "We both looked at each other. I tried establishing communications with the boat to see if it was them, or if the sailboat had capsized."

While Ballard worked the radio, DeMello and Sowder took the government vehicle up to the hill to see if the master chief had seen or heard anything. Zandra would later remember, "Then the *Gale Runner* came back on the air. We then knew what had happened."

Ben Wingo jumped aboard the ready boat last, since he had some problems getting into his exposure suit. Once aboard the 44-foot motor lifeboat, he stowed his SAR bag in the forward compartment. Ben then broke out the pyro vests and handed them to each member of the boat crew.

SN Miniken handed out the surfbelts. Ben and Clint clipped their safety belts into the D rings of the boat. Ben presumed that MK3 Matthew E. Schlimme also had snapped his surfbelt into a D ring, as "he was always so safety conscious." Later, Wingo would not be certain if Bosley wore his safety belt. No one put on his helmet. Wingo said nothing about this.

As the crew of the first lifeboat prepared to get under way, Master Chief LaForge drove to the station. On the way, he stopped near the location of the old U.S. Coast Guard station, now the Quileute tribal school. The location, known as "the hill," affords a good point to observe the bar and surrounding area. LaForge noted the "reduced visibility, but could still make out the light from the sea buoy," about a mile from the bar. He estimated the wind "at 30 knots, with stronger gusts." Rain pelted the area off and on. LaForge also observed "small breaks" on the beach.

As LaForge continued on his way to the station, he spotted the motor lifeboat "leaving the boat basin," but could not recall seeing its running lights. Upon his arrival at the station, Master Chief LaForge radioed the motor lifeboat that he had observed no sailboat in the immediate area and ordered Bosley "to check out the bar to see if they could cross."[3] Communications still had not been reestablished with the *Gale Runner.*

I was jolted awake by the SAR alarm blaring off the barracks wall, followed by the announcement: "Sailboat on the bar taking on water!" I waited a few minutes for the ready boat crew to get to their motor lifeboat. When an SAR alarm rings and you are not a crew member, it is always best to stay out of the way as the crew runs toward the docks. I made my way to the operations room. Near operations, I looked out the double glass entry doors and saw trees whipping in the wind and driving rain. Unlike in movies or television, radio transmissions between sea and shore are never crystal clear, especially when a storm strikes in the midst of an SAR case. Transmissions are usually bad: static and other stations compete with stress-filled voices. Adrenaline surges and voices rise, which make transmissions even more difficult to understand.

When I arrived outside the communications room, DeMello, Ballard, and Fe Brantley were all handling the radio, intercom, and telephone traffic and messages from Group Port Angeles (fig. 9.5). Master Chief LaForge, after talking on the radio, had changed into his exposure suit and come back to the communications room. BM1 Placido had also arrived and donned his dry suit as he listened to the reports filtering into the communications room.

Fig. 9.5. SA Falicia Brantley. Photo courtesy Falicia (Brantley) DeMello.

The Group advised the station to keep the 44-foot motor lifeboat from crossing the bar, always a hazardous and dangerous operation in high seas and limited visibility, because the call might be a hoax. The watchstanders began calling the motor lifeboat. They received a "we're busy" from Bosley.

LaForge could tell from the tone of Bosley's "we're busy" that he was perhaps crossing the bar, and "this was not a good time to distract him with radio traffic." LaForge told the watchstander he would go on the second boat as soon as it had a full crew. Placido said he was going to go. The master chief agreed, and a few minutes later LaForge said to me, "Come on with me, Dennis, while I check the bar." I told him to wait a

Fig. 9.6. Rugged James Island. Photo by Dennis L. Noble.

minute while I ran to my room to get a coat. A few minutes later we were heading toward the hill.

We drove in a four-wheel-drive Isuzu equipped with a radio. LaForge heard Bosley say in a more reassured voice that they were across the bar and the seas were running 16 to 18 feet, but things were evening out. LaForge felt that Bosley had turned the motor lifeboat away from James Island and was headed toward the sea buoy and into deep water (fig. 9.6).

LaForge and I arrived at the observation point and peered through the driving rain. The master chief could not make out the navigation lights of the ready motor lifeboat. I heard a faint transmission that sounded like: "We rolled the boat." LaForge and I discussed what the transmission sounded like. LaForge tried to raise Bosley's motor lifeboat. He heard "disoriented." LaForge saw a spotlight pointed toward the south, from what looked like "the seaward side of James Island." Glimpsed only momentarily, the beam swept rapidly through the night. George knew the ready motor lifeboat was in trouble and again told the station to get the second boat under way. He also told the station to have Port Angeles launch a helicopter. The station called LaForge; the duty officer at Port

Angeles wanted to talk to him directly. The officer still felt this might be a hoax.

Suspicion of a hoax was not preposterous. For some unfathomable reason, there are those who think it is amusing to make a false distress call to a rescue organization. During the period from 1994 to 1996, the U.S. Coast Guard received 729 hoaxes. According to Commander Raymond J. Miller, the Group operations officer, there had been a rash of hoax calls during the previous two weeks. Furthermore, TC3 Marshall had informed the duty officer that the voice on the radio sounded like a child. On the recorded transmissions, Infante's voice comes across as a young person.

When Master Chief LaForge and I returned to the station, the Master Chief told BM1 Placido that the ready motor lifeboat was in trouble and might have rolled. LaForge called the Group and told them he had "a serious situation and still needed a HH-65A helicopter launched and again requested the launch of an HH-60." The HH-60 *Jayhawk* would have to be sent from the U.S. Coast Guard Air Station Astoria, Oregon, over a hundred miles to the south.

We will never know what BM2 Bosley had in mind when he pushed the throttles of the ready motor lifeboat forward and maneuvered out into the river, as he did not brief his crew. If Bosley thought the sailboat was on the bar, this would account for his haste in getting under way. In retrospect, listening to the tapes of the radio transmissions of the *Gale Runner,* it is easy to see why Bosley might have made this conclusion. His belief may also account for how he decided to make his trackline after crossing the bar.

Wingo, aboard the ready motor lifeboat, recalls the weather as "rainy and nasty. I knew it was supposed to get bad that night. It was raining so hard we were soaked before we even got on the ocean. . . . I was, like, 'Oh, well, it's raining, *again.*'"

Once under way, Schlimme came over to the starboard side of the coxswain flat and adjusted the radar, then returned to his normal position to the left of Bosley. Miniken operated the spotlight on the port side.

Fig. 9.7. Wash Rock to the left. The steepness of James Island is apparent. Photo by Dennis L. Noble.

Bosley ordered Wingo to use the starboard spotlight in order to light up Wash Rock, near the bar (fig. 9.7).

Wingo felt excited. This was his first night case, and it also meant he would get the rest of the next day off and could sleep in. Later, Wingo recalled Bosley as being impatient. When Schlimme reminded Bosley to be careful that the oil pressure didn't get too high, Bosley said, "Yeah, yeah, I got it."

Somewhere before they reached Wash Rock, Schlimme said to Bosley, "Let's get the fuck out of here!" Bosley said something that sounded like: "Fuck that!" Schlimme may have wanted Bosley to return to the station, or he may have wanted him to move to deeper water (fig. 9.8).

By the time they reached Wash Rock, Wingo, dripping wet, had never seen so much water go *over* the boat as it plunged into the swells. Bosley told Wingo to put the light on James Island; he wanted the island on the starboard beam. Wingo recalls that the island seemed 100 to 150 yards in the distance.

The ready motor lifeboat crossed the bar. Someone said that the Group wanted them to hold off because this might be a hoax. Bosley replied, "I hope not!"

Once across the bar, Bosley called in that they were across the bar and that the seas were 15 to 16 feet and getting better as they went. Wingo later said, "I was about to call BS. Those waves were a lot higher than 15 or 16 feet and it wasn't getting better, but I was, like, 'Oh, well.'[4]

Ben would later recall, "The waves were huge—they would lift us up and I would count, one-thousand-one, one-thousand-two, and then bam! We were up there that long, and you had to bend your knees, or they would shatter.

"The storm was so loud, the wind, the slamming waves, we could barely hear each other shout and we were standing three feet apart. It was so dark we couldn't see the bow of the boat."[5]

Much later, Ben Wingo would remember, "If I'd known then what I know now, I would have known it was really serious, but to tell you the truth, I was having fun. I thought "Wow! All right! You know how you get that [wooo] feeling."[6]

Fig. 9.8. MK2 Matthew E. Schlimme at work on a Quillayute River station 44-foot motor lifeboat. Photo courtesy Thomas L. Byrd.

The boat moved along the island. Wingo spotted a rock and yelled to Bosley, "Rock! Starboard side! Ten feet!"

The boat hit something. Bosley yelled, "What was that?"

A shout: "Wave, port side!"

Wingo turned his head and spotted an enormous wall of white.

"All of a sudden we were underwater. I hit my face on something and shattered my nose something fierce. I could taste the blood and the salt water. It was like we were under for a long time. I was spinning, I guess, as the boat rolled because when we came up, my surf belt was all tangled and it had been straight before."[7]

When the 44-footer came up, the bow pointed toward James Island. Bosley grabbed the radio microphone and called that they had rolled the boat and the crew was disoriented. Wingo found himself wrapped in a canvas dodger, used to help protect the crew in foul weather. He quickly

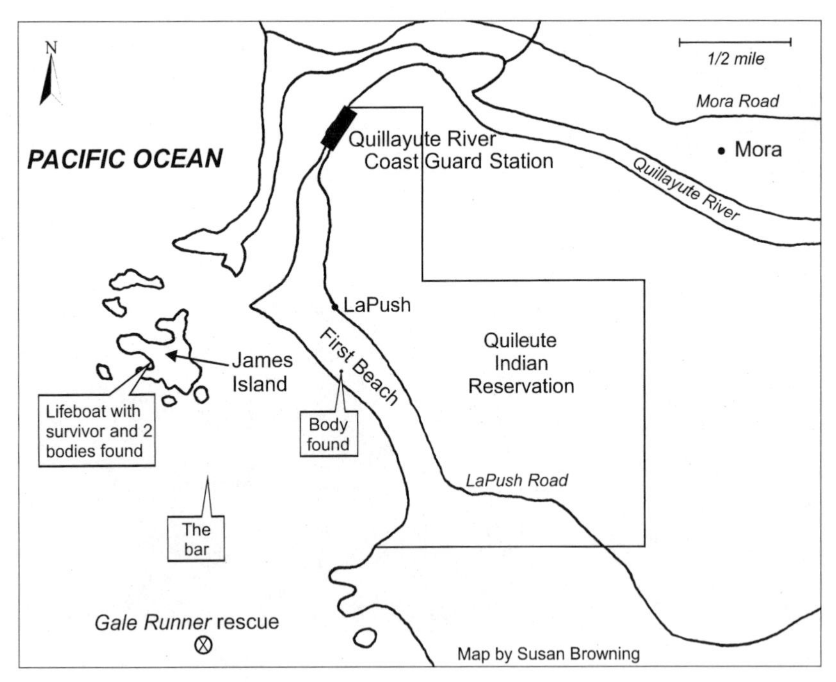

Map 3. The coast near LaPush, Washington. All ocean areas on this map are considered the bar.

Fig. 9.9. SN Clinton P. Miniken upon graduation from Snohomish, Washington, High School in 1992. Photo courtesy Daniel Miniken.

freed himself. Schlimme said they were still on the bar. Wingo reached for the spotlight. Gone. Bosley yelled, "Look for buoy 3." Wingo could see the lights of the village of LaPush, but not buoy number 3.[8]

Another shout. The boat pitchpoled. When they came up, the boat rested on rocks. Ben could tell by the engine noise that the 44-footer was out of the water. The entire top of the forward cabin was gone, as was the mast. So were Bosley—the man who loved to drive boats and to fish—and Miniken, a hard-working kid who always tried to do his best (fig. 9.9).

MK3 Schlimme now took command of the boat. Wingo panicked and started to unbuckle. "We've got to get out of here!" Schlimme made Wingo get back into his belt. He reminded Wingo they had to stay with the boat.

Schlimme asked Ben for the radio. Ben heard Schlimme call in their position and situation, then looked down and saw that the radio was broken. Later, Wingo would recall, "He was just doing it to calm me down, and it worked because suddenly I was super calm.

Fig. 9.10. The 44363 inside the cove. The top of the cabin has been sheared off. Photograph courtesy PA3 Della Price, U.S. Coast Guard.

"He was an awesome guy, an awesome engineer, and he really knew his stuff. He saved my life by keeping me on the boat."[9]

They needed to get inside the boat, but the hatch to the forward compartment was jammed. Then Schlimme yelled, "Hold on!" A wave knocked the boat off the rock and rolled the boat for the third time.

Below the surface there was a "tremendous quiet," Ben recalled. "I don't know how to explain it. The night is raging. There's all this noise, the wind, the water, and suddenly it's just absolutely quiet."[10]

When the boat came up, Schlimme—the kid with the ready smile—was gone. Wingo was alone. He saw a body go by the boat, but could not throw a life ring, as both life rings were gone.

Wingo realized the motor lifeboat was moving backward into the cove on the island. The boat banged along the rocks. He turned on the strobe light in his pyro vest.

Ben looked at his watch. It was 1:07 A.M.. He took the flares from his pyro vest and shot five into the air. Ben thinks he put the engine throttles into neutral. He attempted to get into the forward cabin, hoping to get the emergency radio and battle lantern. The hatch could not be opened.

He thought: "What can I do? What can I do? I'm not the most devoted servant but I have Christian beliefs. I started praying. It wasn't a prayer of all the things I would do or wouldn't do if I was saved, it was just an all-out cry for help: 'Please get my boat to shore.'"[11] He used his last two flares to help illuminate the beach area. The boat swung stern to the beach.

Then the night closed in again. Ben strained to see through the darkness. He saw what he thought to be the dim form of a tree. He unclipped his safety belt and leaped from the boat. He landed in knee-deep water. With a combination of fear and adrenaline pumping, Ben reached the beach, activated a personal marker light, and scrambled up a cliff to await rescue.

Bosley, Schlimme, and Miniken were dead by the time the Port Angeles helicopter got off the ground. The rollovers happened that fast (figs. 9.10 and 9.11).

Fig. 9.11. CG 44363 in the cove. Note the size of the drift logs. Photograph courtesy PA3 Della Price, U.S. Coast Guard.

. . . 10 "I'm going!"

While Ben Wingo clung to a cliff on James Island, the second boat crew was recalled to the station. BM3 Marcus M. Martin remembered, "The telephone rang. [If] the telephone rings after 10:00 P.M., I automatically think SAR case. My wife, Terise, picked up the phone. She said, 'Hello. Oh, really?' I was up and to the closet and before she hung up the phone I was dressed. I gave her a kiss and left. With the weather, I knew something was going on. From the time the phone rang until I reached the station was about three minutes."

Two years later, in July 1999, Marcus told me his experiences aboard the second boat that long night in February. Interestingly, on this July evening his voice held the traces of the fatigue he was fighting after a 14-hour lifesaving mission in a storm.

"At the station they told us to get dressed. I went into the duty room. Stoudenmire and Petty Officer Byrd were there. I'm putting on my . . . [exposure coveralls]. Byrd says, 'It's nasty out there. It's really nasty out there.' So Stoudenmire and I took off our . . . [coveralls] and put on our dry suits. I later thanked Byrd for that warning.

"When we walked out, Jon was walking in. He said, 'Go down to the boat.' So we knew we were getting under way.

"As I was passing the communications room, I thought I heard 'capsized' and thought they were talking about the sailboat. I thought they were talking about another boat other than our own."

Master Chief LaForge and I returned to the station. LaForge called the duty officer at Port Angeles, 70 miles away, and told him he had a serious situation, that he still needed a HH-65A helicopter; he again requested the larger HH-60 from Astoria, Oregon. I watched both LaForge and Placido becoming uneasy (fig. 10.1).

"Flares!"

"What color?" asked LaForge.

"Red."

"I'm going!" Placido shouted, as he ran toward the second 44-foot motor lifeboat.

"Let's go back to the bar," LaForge said to me.

Parking again at our observation point on the hill, I could see a red flare arcing through the sky. I have participated in a number of cases involving flare sightings, all of which proved to be false. This was the first instance in which I actually saw one fired in distress.

"We got down to the boat and started doing our boat checks and getting everything ready to go," said John Stoudenmire. "We got the helmets, vests, and surf belts out. Jon Placido came down and we got a quick brief, but at that point we didn't know our boat was in any distress."

Placido arrived at the standby motor lifeboat, the CG 44393, and found his crew, made up of MK2 Thomas L. Byrd, BM3 Marcus M. Martin, and SN John A. Stoudenmire III, waiting (fig. 10.2). In another of those "what-ifs" that make up this incident, BM3 Martin's wife, Terise, was having a difficult first pregnancy. As a break-in coxswain, Martin should have been on the first boat, but LaForge and Placido tried to give Martin as much time with his wife as possible, so they assigned him to second boat. Two years later, Marcus would tell me, "I look at my daughter and think, 'I wouldn't be alive if it wasn't for her.'"[1]

Placido briefed his crew. "I originally wanted to get out of the river and stick my nose close to the bar and look around before trying to get across. I had no Goddamn clue what the bar was doing."

Fig. 10.1. BM1 Jonathan A. Placido (*left*) on the mess deck of the Quillayute River station with BMCM George A. LaForge standing next to him. Photo courtesy Thomas L. Byrd.

Just after they left the boat basin, however, Jon saw a red flare arcing through the sky. He felt this might be from the ready motor lifeboat and briefed his crew about his concerns as the second motor lifeboat moved down river toward the sea. The sighting of the flare then determined the course of action: Placido had to cross the bar. After determining the conditions, he pushed the throttles forward and turned to cross the bar.

Master Chief LaForge and I had again returned to the hill. LaForge could help judge whether Placido was being set toward rocky James Island.

Suddenly a bright light moved through the darkness, up, down, sideways, indicating a roiled sea. The 44-footer with Petty Officer Placido and his crew of three came into sight. The boat pitched and rolled, as only a motor lifeboat can in a heavy sea. It turned from the protection of the river to make its run across the bar.

I will always carry with me the sight of Petty Officer Placido's 44-foot motor lifeboat as it entered a wave on its passage—waves that were later estimated to be at least 16 to 18 feet. A searchlight probed the dark and the waters ahead, looking for obstacles. Small white boat rising. Rising. Rising. Rising until it seemed to stand on its stern. White water almost enveloping the small boat. Then the plunge downward.

Much later, when I talked to Jon Placido about this awful night, he tried to play the macho surfman: "I wasn't scared," implying that no sea was going to worry him. Later still, when we had visited longer, he finally admitted to me: "When that first wave hit, I was scared." I was too, and I was not even in the boat.

Master Chief LaForge, using the radio in the Isuzu, kept up a conversation with Placido's boat. Meanwhile, Group Port Angeles now classified the call from the sailboat as definitely not a hoax. The station had

Fig. 10.2. *Left to right*: SN John A. Stoudenmire III, MK2 Thomas L. Byrd, and SA Benjamin F. Wingo under way on a Quillayute River station 44-foot motor lifeboat. Photo courtesy Thomas L. Byrd.

received radio transmissions from a frightened woman who was having a hard time understanding the Quillayute River station's instructions, complicating a situation that was becoming more and more worrisome.

Master Chief LaForge, watching Placido's boat, could tell by how the motor lifeboat's navigation lights were moving and the time interval it took for them to bob up and down that the swells were lengthening out as the motor lifeboat got further out to sea, thus reducing some of the danger they posed. (The closer the waves are together, with their constant pounding, the more dangerous they are to a small boat.) LaForge also thought that the second motor lifeboat seemed to be setting northward—toward James Island—and he called Placido to let him know. He received a "roger that," and the motor lifeboat turned and headed toward the deeper water by the sea buoy.

More flares arched from the right to the left of our field of vision, from somewhere in the vicinity of the western part of James Island. LaForge now knew Bosley's crew was in deep trouble. Furthermore, the rocky nature of the terrain, plus the reflective waves, would make it impossible for Placido to rescue anyone.

"The whole passage down the river is a blur to me," recalled Marcus Martin. "The first thing I remember I was on the spotlight trying to find Wash Rock.

"Jon said to me, 'Find Wash Rock!'

"In the rain and all that, I could not find Wash Rock. We were taking breaks and that spotlight was going everywhere. I'm hanging on, sticking my head out, just getting pelted by the rain. I'll never forget. Jon goes, 'Marcus!'—he never called me by my first name, so it's gotta be important—'*Marcus! You gotta find it! I can't turn up until you find it!*'

"I never saw Wash Rock. I guess I lost my balance and I turned and saw the silhouette of James Island. So, I said to myself, 'Okay, James Island, then 20 feet—30 feet—Wash Rock. We're past it.' I guessed. I said, 'Turn up.' So, when Jon said 'We're turning up.' I was holdin' my breath a little bit.

"When we turned up, I was still lookin' for Wash Rock. I knew if we hit it, it would be right down my side. I can't find it. I can't find it. Then I knew we were past it. So, there's a little bit of relief."

"We had our spotlight on," Tom Byrd recalled. "I never will forget seeing those waves breaking over that bar. I was thinking to myself, 'Oh, my gosh! I can't believe we've got to go into this!'

"The waves were huge. Knowing once we got out we're not going to be coming in—when you go out in something like that at night you can't come in over the bar at night, plus the bar lights were out. It was pitch dark, pouring down rain.

"At the mouth of the river you could see these giant waves going by. Placido was waiting, trying to time it just right between the series of waves. Soon as we got out, we turned into the waves and we hit.

"The waves and the rain made it so that we couldn't see anything on the radar. [Master Chief LaForge] up on the hill helped us a lot [by radioing] us about the waves and the direction to go. I was about as scared as I have ever been in the 20 years I was in the service. I, of course, was the first one to throw up. Stoudenmire was on my left, behind me. I threw up and it went all over him. He said, 'That's all right Petty Officer Byrd, you can throw up.' A few minutes later everyone was doing it, so I didn't feel so bad.

"You can't tell someone how rough it was, it is something you have to experience. We were just pounded, pounded, pounded."

"I was looking for breakers," said John Stoudenmire. "It was so dark you couldn't see them until they were right upon the boat. I mean, right at the bow. That was pretty nerve-racking, but you are so pumped-up—at least I was—I really wasn't scared, at least not as much as I was later."

"I'm trying to look at the island," recalled Marcus. "The next thing I know we're in the air. I heard the engines. I heard the props turnin.' I looked over and Tom's already sick. He's on the deck. We go up in the air and then, Tom's at eye level. His whole body is in the air. Stoudenmire is hangin' on to him. I'm hanging onto the 'oh, shit bar.' My back is pressed up against the roof. We were 44-feet of pure metal and steel going straight down in zero gravity.

"We hit. BANG!

"We went a little while. Then got hit by another one. I think we got hit by three big ones. It was just the reflection off the island swell. We were in that washing machine, 'cause Jon was hugging the island.

"Master chief called us saying he lost us, we're getting too close to the island. We got hit by the last one, which I think was the biggest. I'm not sure, because they all were big. Jon says to the Master Chief, 'Roger, I'm cutting out.'

"The antenna broke off. There was so much water it shorted out our radio. We went down into the forward compartment and got our handheld. Eventually that died."

"My crew performed for me like they were supposed to," Jon Placido related. "They constantly fed me information. I felt like a machine. I absorbed the information and my body performed the necessary motions." Placido did not mention the skill and experience that enabled him to take the boat across a bar in horrible conditions and afterward bring his crew back safely.

"When I first did the corner around the bar and we were heading out, I was pretty comfortable," Jon recalled. "When we got hit by those big waves, I had one thing in mind: survival. I knew that I had to get to the entrance buoy. Once I was out to the buoy, and my sense of self-preservation kind of died down, I realized now I gotta do somethin'.

"We now had flares to the south and to the north. When we were looking at the flares, I knew which ones were ours. I started to push back in and had to decide: Should I go for the sailboat, or our boat? It was one of the hardest decisions I ever had to make.

"I saw the helicopter starting to search for our boat. I knew the helo could probably help out our boat better than I could, so I headed for the sailboat.

Everyone lost communications with Petty Officer Placido's boat. Calls from Master Chief LaForge and the station went unanswered.

A woman's nervous voice. "Are you coming out to help us?"

LaForge again called for helicopter assistance, which would entail a long flight from Port Angeles in heavy wind.

More calls. Now Master Chief LaForge had a sailboat in distress with two people aboard, a missing motor lifeboat with four people aboard, and another boat with four people aboard not responding to radio calls.

Finally, Petty Officer Placido: "We are on hand-held [radio]. Our antenna was damaged by a breaker." Master Chief LaForge advised Placido to go to the "Q" buoy and remain in deep water. Placido and his crew would linger by the buoy until daylight, at least six hours of fighting the sea in the 44-footer.

LaForge and I returned to the station.

Back in the communications room at Quillayute River, Zandra Ballard remembers that when the flares were shot they seemed to come "almost one after the other from James Island. There was so much going on. We started to recall the rest of the crew. Not one person questioned what was going on, or showed any sort of anger that they were being called in. This was part of their family. My personal feeling at this time was that the crew of 44363 [the ready motor lifeboat] was fine and that they were just stuck. I guess I didn't think that one could roll on the boat and then die."

By the time LaForge and I arrived back at the station, communications with the sailboat were now well established. The *Gale Runner*'s position was fixed at a location near The Needles, the jagged rocks sticking out of the ocean approximately three miles to the south of Quillayute River. Master Chief LaForge felt this position was consistent with the wind direction and speed. LaForge asked if the sailboat had propulsion. Back came a short "no." Master Chief LaForge, who is also a sailing enthusiast, asked if they could sail to seaward using bare poles.

Infante responded with a disheartening, "No, we're dismasted."

A shaking of heads in the communication center.

Master Chief LaForge ordered a beach search in the vicinity. The Quillayute River crew would wear their exposure coveralls and helmets. National Park Service rangers from the Mora Ranger Station, about 17 road miles away, joined in the search. John DeMello volunteered to go. Eventually John, MK3 James C. "Scooter" Johnson, and FN Matthews were detailed to the search, with MK1 Mumford driving them to the scene. "I think it was automatically decided that me staying in the comms room was best," recalled Zandra Ballard.

The first helicopter, number 6589, from Port Angeles, arrived in the vicinity of James Island and began the search for Bosley and his crew aboard the ready motor lifeboat. Master Chief LaForge now had to made an agonizing decision. He knew his motor lifeboat crew was in serious trouble and missing, but he also had two people on a sailboat drifting inexorably toward the rocks and sure death. Master Chief LaForge called the helicopter and recommended that it break off the search and head for the *Gale Runner.* Zandra Ballard would later say, "I don't know what was going through Master Chief's head, but the look on his face was one that I will never forget.

Ballard would also remember that when the Port Angeles helicopter began to transmit their prehoist checklist, the communications were so bad the pilot asked the station to transmit them to the *Gale Runner.* Master Chief LaForge called the *Gale Runner* and told them to prepare for a helicopter hoist and transmitted the required instructions. Infante came over the radio asking if she could bring her cat with her. "It's your decision," said LaForge, "but the cat will have to be in a bag."

I recall Master Chief LaForge turning to the radio watchstander and saying, "Keep off the radios as much as possible. The pilot may have only one short chance to broadcast if he goes down." Within two and a half-hours of being awakened by the ringing of his telephone on a stormy night, LaForge had 14 lives, two motor lifeboats, and a helicopter as his responsibility, with a very good chance of losing at least 10 of those 14 lives from the sailboat, one motor lifeboat, and a helicopter.

11 "You have 60 seconds to prepare yourselves"

Quartermaster First Class (QM1) Joseph N. Sekerak, the Group duty officer in Port Angeles, was asleep in the duty bunkroom when he was awakened by a telephone call from TC3 Gina Marshall informing him that she had heard a Mayday and that "it sounded like a kid." Sekerak came into the operations center, and TC3 Marshall briefed him about the Mayday call. Later, Sekerak would remember, "I did not know that . . . Quillayute River was launching their motor lifeboat or that it was under way." Sekerak had noticed on his way to the operations center that Commander Raymond J. Miller was still working in his office. The Group duty officer then went to Commander Miller's office, briefed him, and asked him to step into the operations center.

When Commander Miller arrived in the operations center, he asked about the weather at Quillayute River. Sekerak passed on the weather. Neither man knew that Bosley had already decided to launch the ready motor lifeboat. Sekerak said the call could be a hoax. Marshall then said, "although the voice sounded young, it was very panicked, and I estimated the sailing vessel to be between Quillayute River and Grays Harbor."

Commander Miller would later recall the analysis he made of the situation. "The initial bar report request call to station Quillayute River had been loud and clear, but the second call was broken and barely readable. This, combined with the voice of the caller becoming noticeably shaky on the second call after a relatively calm first call, told me that the vessel may have sustained enough damage to dismast it. Second, and just as important to my belief that the calls were the real thing, was that the calls were coming in on the James Island VHF-FM high site and were also heard by station Quillayute River's watchstander on their low site, indicating the calls were coming from somewhere off the Washington Pacific Coast. There had been a series of hoax calls from a person, who sounded like a young male, over the past several weeks, and the voice on these calls indeed did sound quite similar to the voice the GDO and TCOW [telecommunications technician of the watch—Gina Marshall] were hearing, but all of those hoax calls had been received on a high site in the San Juan Islands, nearly 100 miles northeast of James Island, and on the low site at station Bellingham [Washington]. The GDO and TCOW agreed when I summarized my view."

Commander Miller "told the GDO to call the SDO and Quillayute River and have them launch their motor lifeboat. I reminded the GDO to call the commanding officer." Miller then changed into his dry suit and, before he could get out the door, Sekerak informed him that Quillayute River had launched their motor lifeboat.

Sometime before 1:00 A.M., Commander Paul A. Langlois came awake to the ringing of the telephone in the Senior duty officer's quarters. The Group duty officer informed him of a Mayday from a sailboat off LaPush. "I was in a pretty deep sleep when I got the call," recalled Langlois. "I told the GDO to hit the SAR alarm and provide the rescue swimmer and night-vision goggles. He said that station Quillayute River was getting a boat underway, and that the position of the sailboat was several miles south of the station. I got into my dry suit and headed down to see the GDO before going over to the hanger."

Captain Philip C. Volk later said, "Because the XO and OPS were standing the duty, I took the calls. That is, when a helicopter launches on a nonroutine event the operations officer is called and told what is going on. He either comes in to help or takes calls from home. If the OPS officer has the duty, the XO takes the calls, but when they both have the duty, then the CO takes the calls.

"I got a call from QM1 Sekerak, who told me they had received a radio call that was somewhat garbled, no position, the word 'Mayday' was somewhere in the radio call, but they didn't know who it was, or where it was, and they believed it could be a hoax. The weather was so incredibly foul they honestly didn't believe someone was out there. They were evaluating it to find out what was going on. Petty Officer Sekerak said the operations officer had been in the operations center, and the XO had been in bed but had come down to the center. Nobody was going anywhere until they had evaluated what they had heard on the radio. I felt good about that. They were evaluating everything; the weather was terrible, but no one was going anywhere. It was unbelievable someone would be out there.

"What he didn't tell me at the time, and he may not have known about it, was the earlier call the *Gale Runner* had made about the bar and bar conditions. I always ask what is the weather like. The weather was pretty crummy. He reassured me no one was going anywhere. He would call me back.

"I went back to bed.

"The second phone call he told me they had lost comms [communications] with the 44-footer, they had spotted red flares, the helicopter was launching, and that I needed to come in. I dressed in a matter of seconds and got in the car and drove out to the air station. I set some kind of record for driving in.

"Gina, the TC, was on duty, cool, calm, and collected, as was Sekerak. Shortly after I got there the JOOD got there, a young seaman. This was standard operating procedure. Whenever a case happened, the JOOD was the note taker. I told the JOOD, 'You sit here and you write what we tell you to write and when we don't tell you to write, write what you see and hear. You write times and don't answer the telephone, don't talk to anyone, just write.' I got briefed on what we knew and it was not until then that I knew the 44-footer had got under way.

"When the helicopter is launched, the Group takes senior mission controller, and because of that, contrary to popular opinion, I was the senior mission controller. The on-scene commander was the station [Quillayute River]; until the helicopter got there, the group ran the case; the district was notified and was briefed."

Captain Volk's long night had just begun.

ASM1 Chuck Carter said, "I'd just gotten into that heavy sleep and I came awake [to the SAR alarm] and I was mad. We hadn't had any mentionable SAR cases for so long and you think flare sighting and that amounts to nothing. I felt like I could punch someone. Coming down the stairs, I could hear someone say, 'Well, we think it's a hoax.' So that added to it. I was like a little kid.

"We get to the helicopter and I'm kinda griping and grumbling. I remember suiting up.

"As the helicopter is turning up, the radio call came that it was real and they'd lost comm[unication]s on both 44s. Like, whoa! I was instantly awake. My attitude was completely changed. I went from thinking this is a hoax, to thinking I had *10* people in the water.

"Commander Miller over the radio was going: 'Call the commanding officer. Call Astoria.' He made a bunch of other commands, told the ops center to do this stuff, and then he said, 'Gentlemen, this is the real thing!'

"Instantly I went from being half asleep and miserable to, like, Wow!, what are we going to do here?"

Commander Langlois later said that when the Group duty officer "said something about the boat being dismasted, I then knew the people on the sailboat were definitely in trouble." Paul Langlois then gave his crew, made up of Commander Miller, AM3 Neal Amos, and ASM1 Carter, a briefing of what he knew. Later, Paul would recall, "I knew Ray was one of the Coast Guard's experts with the use of the night-vision goggles and certainly the most experienced pilot at our air station and felt it was smartest to have him wearing them, with myself in the right seat." The helicopter, number 6589, departed Port Angeles at 1:18 A.M., after a short

Fig. 11.1. U.S. Coast Guard helicopter HH-65A, number 6589, at the U.S. Coast Guard Air Station, Port Angeles, Washington. Photo by Dennis L. Noble.

delay due to erratic behavior of the avionics/navigation system (fig. 11.1). Before lifting off, the air crew learned that neither of the motor lifeboats from Quillayute River had reported their positions for at least the past 20 minutes.

Langlois and Miller discussed their flight path for reaching the Quillayute River station area. The helicopter began to run into low clouds and moderate rain, with strong southwest winds. "Flying direct to the scene," recalled Langlois, "over the mountains would have been a shorter distance, but was too dangerous due to potential icing conditions, probable turbulence, and high terrain in instrument conditions." Instead, the pilots elected to fly along the Strait of Juan de Fuca to Tatoosh Island and then turn south along the coast to the Quillayute River area.

Commander Langlois flew the helicopter between 500 to 800 feet off the water. Cockpit instruments registered a temperature of 34°F and winds from the southwest at 40 knots during the leg from Port Angeles to Tatoosh Island. "We were picking up a little ice on the helicopter and gusts were up to 60 knots," recalled Langlois. "We took quite a beating over the strait. The winds were from the southwest and they tend to spin off the mountains to cause turbulence and that causes a rough ride, which causes the pucker factor to go up." The weather necessitated an instrument flight all the way to the Quillayute River area. "We flew at maximum performance," said Langlois, "which gave us about 140 knots indicated airspeed."

Commander Langlois flew the helicopter, while Commander Miller quickly became comfortable with the night-vision goggles, handled the communications, and backed up the pilot in command with navigation. "Night-vision goggles give about 47 degrees field of view, with a greenish cast. That can be disconcerting. You don't have the sense of depth perception and peripheral vision," said Ray Miller.

"Our attention and focus had shifted from the dismasted sailboat to the grim reality that our own motor lifeboat might be in distress," said Paul Langlois. "We all privately hoped the motor lifeboat had just lost their comms and were otherwise all right."

Turning onto the southbound leg at Tatoosh Island, the winds were 40 knots, with some gusts to 53 knots. The pilots of the helicopter could only pick up weak broadcasts from the second motor lifeboat. Later, they would learn that the weak communications came from the handheld portable FM radio BM1 Jon Placido was forced to use after a breaker damaged the motor lifeboat's antenna.

The pilots heard a "very scared" female voice on the radio give a precise latitude and longitude of their position. "I suspected she had GPS," said Langlois. "I envisioned that we would be able to easily locate the sailboat and not have to search after we first located the motor lifeboat with the lost comms."

"I don't know when we realized how the weather was," said ASM1 Carter. "The night had been kinda nasty, and I don't remember it really hitting me until we were almost out there. On the way, we discussed what we knew and didn't know. As we turned the corner and headed south, we knew one of the 44s was in the water and so when we got on scene we had a question. Do we look for the Coast Guard people in the water, or do we go to the people on the boat? I heard Commander Miller discussing this decision. You think, 'Wow, this is a really tough decision.' You think a commander can make this decision. You know, you got Coast Guard people who might be dying and you got civilian people who might be dying."

Approximately 39 minutes after liftoff from Port Angeles, at 1:57 A.M., the Port Angeles helicopter arrived off James Island. Commander Paul Langlois shifted his view from the instruments to the outside and quickly viewed the lights of LaPush. He spotted the second motor lifeboat, the CG 44393, driven by BM1 Placido, near the Quillayute River sea buoy. Paul later recalled the spotlight of the motor lifeboat, which normally would be tracking at the horizon, "was going in oscillations of almost straight down to vertical. I could tell from Placido's voice that he was not enjoying the situation he was in. I then realized how high the seas must have been."

Commander Paul Langlois descended to approximately 300 feet and slowed to 70 knots. He discussed with Commander Miller about staying on the night-vision goggles, even if they had to descend below 300 feet, despite restrictions in the *Air Operations Manual.* Paul knew the rugged coastline. The wind direction would push the helicopter toward the east and into the rocky spires and cliffs. Ray Miller agreed to keep on the night-vision goggles.

Commander Langlois, as the pilot in command, briefed his crew that they were searching for Quillayute River's ready lifeboat, the CG 44363, before going to the assistance of the sailboat. Paul would fly the aircraft on instruments, while Ray would control the heading with the heading knob located on the instrument panel near his right leg. Most importantly, with the night-vision goggles, he would be able to better locate the rocks.

After several minutes of searching for the motor lifeboat, the pilots

Fig. 11.2. The Needles. The air crew of helicopter 6589 flew around, over, and between these rock formations in a rain-swept gale at night. Photo by Dennis L. Noble.

heard Master Chief LaForge's voice on the radio requesting them to switch to channel 81, the U.S. Coast Guard's working frequency. Once on 81, the master chief recommended they break off the search for the CG 44363, as his plot indicated the sailboat drifting rapidly toward The Needles. Later, Ray Miller would say, "Our entire crew found it very difficult to interrupt our search for the 44363."

Langlois turned the helicopter toward the last reported position of the *Gale Runner.* A few minutes later, the helicopter crew spotted a dim light ahead. Ray informed Paul that through the night-vision goggles he spotted the rock formations near the sailboat. The steep, towering rocks would interfere with their descent to a hover (figs. 11.2 and 11.3).

Langlois informed his crew they would do a basket recovery of the people aboard the *Gale Runner.* The use of the rescue swimmer was too dangerous in the crashing seas among the rocks. Paul informed the other three crew members that he would use instruments to make a letdown to a hover. Ray Miller would stay on the night-vision goggles continuously and give him any steering commands around the rocks, while he attempted to keep the helicopter pointed into the wind. Besides the wind and darkness, the helicopter flew through lashing horizontal rain.

Commander Miller would recall, "I could discern the rock formations of The Needles. I told Commander Langlois that we would have to work our way over and around some high rocks to get down where we needed to be to conduct a hoist." Miller continued on the night-vision goggles and controlled the heading while advising Langlois of the descent rate.

Through the night-vision goggles, Commander Miller could see the *Gale Runner* to the west of the rocks and about 300 yards away from them. Ray knew the helicopter would have to make a steep approach into the wind to safely clear the rocks and arrive at an altitude to hover near the sailboat. "I was comfortable on the night-vision goggles. I'm

Fig. 11.3. The Needles can rise to a height of a 16-story building. Photo by Dennis L. Noble.

not too sure how comfortable the crew was after I described the rocks." Some of the rock spires rose 190 feet into the air.

"I could see the rocks very well," said Miller. The helicopter flew over *and between* the rocks, ending at 150 feet above the *Gale Runner* and at 65 knots indicated airspeed. Too high and fast. Both pilots called for a go-around. The helicopter turned right and departed toward LaPush. During the terror-filled minutes ahead, the four U.S. Coast Guardsmen in the helicopter would be dodging between rocks that rose to at least the height of a 19-story building.

Commander Miller fine-tuned the approach and told Commander Langlois when they had cleared the rocks. Langlois then briefed his crew on a trail-line delivery.

Commander Miller called the sailboat. "I instructed them to clear any downed rigging from the cockpit as best they could. I explained the trail-line delivery of the rescue basket."

Langlois "instructed Petty Officer Carter to keep an active scan out his window to the back, and advise us when he saw the rocks."

Commander Langlois told Miller not to let him get below 25 feet over the wave tops. AM3 Neal Amos began to talk Langlois over the *Gale Runner.* The wild movement of the sailboat and the battering wind prevented the helicopter from hovering over the craft. "Neal certainly was doing his best at giving me conning commands," said Paul. "We were both very frustrated. We couldn't maintain our position in relation to the boat."

ASM1 Chuck Carter recalled, "We're trying to do the basket hoist and it's just not working. Neal said to Commander Langlois, 'You've gotta hold it steady.' Commander Langlois said, 'I am.'

"'*You're not doing it, sir!*'

"It was just frustration. Everybody was frustrated."

Trying to make the hoist work, Paul Langlois *backed* the helicopter through the gale-swept night, closer and closer to the rock pinnacles. "None of us were absolutely sure what obstructions were directly behind us," said Langlois. "My focus continued with trying to get the people off the boat before they hit the rocks."

ASM1 Chuck Carter said, "In the light from the helicopter's searchlight I can remember the boat coming up and dropping down the backside of a wave. It looked like they fell off the face of the earth. You could see them. Then they were gone."

Langlois recalled that AM3 Amos said the trail line was streaming straight back in the wind, despite the weight bags. "I told him to get rid of the trail line. Let's see if we could just put the basket on deck directly." Paul would later say, "Throughout the whole hovering sequence, Ray was giving me incredibly precise advisories on power, altitude, and whatever he could see through his night-vision goggles." Then Langlois suffered from vertigo. Ray helped keep the helicopter steady. Langlois, at one point, shifted his scan to the limited view to the rear of the helicopter. "I saw a large rock pinnacle right behind our aircraft at the five o'clock position. ASM1 Chuck Carter thankfully alerted me to the rock." *The rock was higher than the altitude of the helicopter.* Paul Langlois would later say, "I have to admit I got pretty scared at that point, and now fully realized the danger in which I had placed the whole air crew." Then Commander Langlois related something most people in maritime search and rescue eventually say: "I'm sure I became reliant on adrenaline and instinct to continue the effort." Commander Langlois continued to try to get the basket onto the *Gale Runner,* but to no avail.

ASM1 Chuck Carter said, "'We've gotta do something different.'

"They said, 'What?'

"I said, 'Put me down.'

"'We can't do that. It'll kill you.'

"I really didn't want to go down, but we were at the point of what else are we goin' to do? I said, 'We've gotta do something.'

"At that point we had the basket inside, and Commander Langlois said, 'There's nothing we can do.'

"And I thought, *we're the United States Coast Guard!* If there's nothing we could do, then who could do anything? I remembered I just started praying, 'Lord, if you're out there, you gotta make this happen.'

"And I thought for sure they were going to die. We were going to sit there and watch those people die. The concept that we couldn't help someone never entered my mind. I mean, we can't do anything is not the right answer.

"There was this huge rock. The sailboat was coming at this huge rock. I expected them to hit the rock and they would explode. I thought this boat was just going to be gone."

"I remember this was the most challenging, stressful, frustrating time I've ever had flying—ever!" recalled Paul. "I have always been able to make a rescue; this time I could not do it. As each minute went by, the *Gale Runner* was getting closer and closer to the rocks. Rocks were behind us and higher than the helicopter. I told Ray I could not do this.

"Ray could see that the boat was now approaching some rocks immediately north of The Needles. The surf was getting even bigger, with waves starting to break, completely engulfing the boat. Ray called on the radio and told them they only had a few seconds to get in the cabin and brace for impact with the rocks. Over the next minute, the sailboat was repeatedly swamped in tremendous curling waves, which made the boat disappear from sight for periods of 3 to 5 seconds at a time. Judging from the size of the 31-foot sailboat, I would estimate the breaking waves must have easily been 30 feet. We all privately knew that the people could surely not survive that thrashing. Finally, the boat ended up being tossed on a rock shelf above the surface, laying on its side. By now, I had The Needles directly in front of me, as we had maneuvered clear around it. I remained very concerned that there may be other obstructions right behind us but not in sight."

Ray Miller recalls "telling the people on the boat they had 60 seconds before you crash on the rocks. Go below and prepare yourself." Ken Schlag and Marcia Infante went below.

"I looked through the goggles and could see a narrow gap between the one large spire and the smaller shelf the boat was lying on. As each wave was coming against the windward shore, there was this, like, flushing funnel of water that would go through this narrow gap. Sure enough, their boat was being sucked toward it. As I sat there, I thought to myself, 'Nature has a chance of pulling them through there.'

"Sure enough! The sailboat went right through there. It pitchpoled. It went over and it went under and completely out of sight like a submarine for a short while, for a matter of seconds. It seemed a lot longer to us. Then up it came. After it washed through, breakers were coming through the shelf and the boat got pushed up on another shelf behind the outer rocks and landed on its side. So there it sat, high and dry. A few waves came over it, but not enough to push it over. So we thought. There was a lull between the series. Through the goggles I could see, with the boat on its side, the cabin lights were still on, and I could see people moving in there. I thought, 'They made it through this. They might just be able to climb right out of the cabin. There they are on the shelf and all of a sudden there was this chance for us to scoot over there and pick them up off this rock.'

"No sooner had I thought that and uttered it to the crew, than the next series came through. The boat was on its side with the keel facing toward the sea. The next series came through and got underneath that keel and lifted the whole boat over into deep water on the shore side of the rock and under it went again. It popped right back up. This time it popped up right side up, stern to the waves. It was past three pretty good-sized rocks that [were] moderating the action of the waves, so the boat was still moving around, but it wasn't impossible like it was on the seaward side."

By this time, Paul Langlois had been flying the helicopter in horrible weather for over an hour, including the time to reach The Needles (fig. 11.4). "As worn out as I was at this point," Paul said, "I said to Ray and the flight mechanic, 'I think we might be able to continue to try to do this. Let's keep trying.'

"So very carefully, and here again is where the night-vision goggles paid off, rather than take off in forward flight with the basket hanging

Fig. 11.4. Captain Paul A. Langlois, pilot of helicopter 6589, after his promotion to captain. Photo courtesy U.S. Coast Guard.

out, which you really cannot do, we kind of picked our way, back and around, and got into position right next to the boat," said Ray Miller. "From there on out, it was a matter of only five minutes until we had both people in the helicopter. Paul Langlois did an amazing job after being so drained from all the effort offshore.

"We turned sideways so I could see. I said, 'Come left 50 feet. You're clear of the rocks. Watch out you don't get behind the rocks.'

"With the wind blowing at 40 knots, rocks create a lot of turbulence. You want to stay clear of those zones because that can put you in the water before you can blink.

"We were about 150 to 200 feet from the rocks. It always looks closer when you are in a situation like that." Ray Miller would later remember that throughout the long fight, "I was constantly reminding myself that this is for real. We can do this. It was like there was a self-generated pep rally going on in my head, or a checklist.

"Are we at the outer limits? No. Check!

"Can we do this? Yes. Check!

"Okay, we're still okay.

"I'm still alive. Check!

"They're still alive. Check!

"Every minute or two, this test.

"We actually yanked them off the sailboat. We stressed the hoist so much the mounting brackets had to be replaced."

"Neal did an incredible job of commanding me," recalled Paul. "Neal said the lady is getting into the basket. We dragged the basket down one side of the boat and it snagged. I remember a pretty good tension on the hoist. The basket broke free and came up almost like a pendulum, almost high enough to strike the rotor. I've never seen a basket swing like that before."

ASM1 Chuck Carter remembers looking down at the *Gale Runner* as it shot through the gap in The Needles. "They were completely gone. It was, like, they're gone! They're back! There they are!

"We backed all the way around those needles and I can't really see anything out there. We did the two hoists lickety-split. First she came up and then him. I was so amazed they didn't both come up at the same time. After all we'd been through, I know I wouldn't have waited for another hoist. Come to find out, they were both pretty small people. Weight-wise we could have picked them both up.

"They're in the plane and I'm, like, 'How're you doing?'

"'We're fine.'

"They both looked fine; they weren't bleeding and nothing was broken. I didn't even have time to give them a blanket."

Fig. 11.5. Aircrew of helicopter 6589. *Left to right:* AM2 Neal Amos, ASM1 Charles S. Carter, CDR Raymond J. Miller, and CDR Paul A. Langlois. Amos, Miller, and Langlois received the Distinguished Flying Cross for their work, the highest award for valor in peacetime that aviation personnel in the U.S. Coast Guard can earn. Photo courtesy U.S. Coast Guard.

"We all took a big breath," recalled Commander Langlois. "I said, 'Ray, let's get out of here.' I was completely drained, both mentally and physically. It was the hardest thing I ever did in a helicopter."

The helicopter landed on the Quillayute River station's ball field and an ambulance took the two people from the *Gale Runner* to the hospital in Forks. The Port Angeles air crew flew to the Neah Bay station for fuel (fig. 11.5). On its flight to Neah Bay, another HH-65A, number 6585 from Port Angeles with a recalled crew, passed on the way to the Quillayute River area.

While refueling, the crew listened to reports filtering in on the search for Quillayute River's ready motor lifeboat. CWO2 Robert "Bob" Coster, commanding officer of the Neah Bay station, recalled, "Commander Langlois looked pale and was staring off into the distance. I remember hearing Commander Ray Miller say, 'I can't believe what we've just done!'" At one point during the Neah Bay break, ASM1 Chuck Carter remembered hearing a report of a crewman on the beach, with a dislocated shoulder and being given CPR, and saying, "You don't give CPR for a dislocated shoulder." Later, they learned that an unidentified crewman had indeed been recovered and that CPR was being administered.

The helicopter crew, once again airborne, heard that the Astoria helicopter, HH-60J *Jayhawk* number 6003, had spotted two bodies face down in a cove in James Island. Commander Paul Langlois and his crew began a search pattern, as assigned from Port Angeles, and later heard that the 6003 had spotted a strobe light on the cliff of James Island. "We all hoped that the strobe was coming from a survivor," Langlois said later.

Fig. 11.6. The *Gale Runner* ashore on Second Beach. Note rock formations in background. Photo courtesy Captain Raymond J. Miller.

The first Port Angeles helicopter's crew, completely worn out, again landed at the Quillayute River station's ball field. Commander Langlois went into the station. "I remember the silence and then people crying, hugging each other. Everyone wanted to do something to make it better. Everyone wanted to make it right." Even though fatigued from his ordeal, Paul set about assisting the Quillayute River station's crew, especially with the details of using a high-angle rescue team from the Clallam County Sheriff's Department to check out the strobe light on the cliff of James Island. This marked the ending of role of the crew of helicopter 6589 in aerial operations at Quillayute River.

The public affairs officer for Group Port Angeles has pointed out that since the Quillayute River station fell under the command of Group Port Angeles, it was only natural for Commanders Langlois and Miller to assist the station after their hazardous flight. After interviewing and observing these two pilots, I suspect that both would have offered to help no matter what the chain of command.

Commander Paul A. Langlois would much later say, "We all believed something miraculous helped us. Somebody above was watching over us. There must have been enough pain in losing our own crew that somehow it was right to help those two people."

. . . 12 The Deadly Beach

While BM1 Placido and his crew fought the towering seas near the Quillayute River entrance buoy and Commander Paul Langlois and his helicopter crew moved toward The Needles, Fireman Apprentice John D. DeMello thought Master Chief LaForge had decided to take the RHIB (a small rigid hull inflatable boat powered by outboard engines) to look for the missing boat (fig. 12.1). "Wingo was my best friend on the station. I said I would go. So I ran off to my room to get my . . . [exposure coverall]."

When DeMello returned, the plan had changed. "Someone said, 'We need a beach party.' I was already in my . . . [exposure coverall]. I said, 'I'll volunteer.' So it was Fireman Matthews, Petty Officer Johnston, and myself that would go on the beach patrol. Petty Officer Mumford was the driver.

"So, we drove out down to the north side of First Beach. There was water coming up to the grass. Petty Officer Mumford said, 'No, this is bad.'

"There was a whole bunch of foam up to my waist. I was used to Hawaii's water, so this was new to me. I hadn't seen an ocean act like this. Coming from Hawaii, you never see logs in the surf. I looked down to the right of me and I seen this huge tree, just rolling because the waves was pushin' it. I was thinkin' 'God! That's dangerous!'

Fig. 12.1. FA John D. DeMello was on communications watch at the station when the *Gale Runner* case began. Photo courtesy John D. DeMello.

"I've still got the picture of Wingo in my mind. What're we going to do next? I've got my adrenaline pumping. This is what I was born to do. I was born to go out and save people and be the hero. That's what I always wanted to be.

"Petty Officer Mumford drove down the beach a little more. Down behind Lonesome Creek Store. We drove down there. Petty Officer Mumford said, 'Okay, we'll stop here.'

"He pulled up and there was a lot of foam, but waves were not coming up to the grass. There were places to walk. He sent me out first.

"Petty Officer Mumford said, 'Look, if you feel unsafe come back to the truck.'

"I walked out. There was this big puddle of foam in front of me. I was walking through it and thinking of nothing else. I fell into this hole and went up to my waist in water. I fell in and quickly pulled myself out. I was, like, Jesus! That's when I first got the notion that everything is not what it seems to be. There's a lot of hidden things here.

"So, I went back to the truck. I was scared. I collected myself and said, 'Yeah, I don't know about goin'."

"We stayed in the truck for about five more minutes. Talkin'. All four of us.

"Will [Matthews] said, 'Okay, let's go and try.'

"I said, 'Let's go as a group.' I was by myself the first time. I didn't want to do that again. Petty Officer Johnston had the hand-held radio. Will had the flashlight, and I had nothing in my hands.

"We started down the beach. I remember there were logs strewn along the beach. We had to run by the water. We had to time it: receding, receding. I counted on Will. He was here a long time before I was, so he knew a lot more about this water.

"Will said, 'Okay, ready! Ready! Follow the wave out! Run! Run!'

"We ran around a log. My heart was racing. We came close. I know those waves were big, but when I looked out all I saw were small waves comin' up on the beach about knee high at most. They had already broke, so it was white wash.

"I ended up getting pretty close to First Beach. I saw a flashlight. So, I'm thinkin' crewman. I started yelling, 'Wingo! Clint!' Nobody answered.

"The light is still flashin', so I thought someone is disoriented and scared. I take off runnin'. Petty Officer Johnston is running after me and so is Will.

"The next thing I hear is, 'Oh, my God!'

"I turned to my right. I saw a log being carried in this wave that is about six feet. The log hit Petty Officer Johnston in the head. He twirled to the ground. Will was a little north of Scooter [Johnston]. He turned to me and said, 'Run!'

"I started running up the beach. The wave caught me. I fell down. The log was lying on my back. I was underwater. I didn't know that until I tried to push up. I couldn't. I felt the log across my back. I'm, like, Jesus, I'm trapped underwater.

"To my amazement that was the calmest moment of my entire life. I've never had something so calm happen to me.

"I was underwater. I remember I looked up. It was really foggy. It was really cold. Then pictures came into my mind. Like my life flashed in front of me. I saw my sister's face. I saw my mom's face. I saw her smile."

"I said to myself, 'All right, I'm running out of breath. I gotta take a breath.' In my mind, I was, like, bye, mom. I started to breath in and got a mouthful of water.

"Then the water receded. I got air. With the water receding, the log rolled off me.

"I got up screaming. I was completely terrified at what had just happened.

"Will was screaming, 'John! Scooter! John! Scooter!'

"He was way north of me. The wave must of carried him that far. He was running back toward me. I got up. He grabbed me. I said, 'My arm hurts.'

"They said my hand was hanging below my knee. My shoulder was dislocated. I found out later it was broken. It was hurting a little bit, but I was in shock, so I didn't feel it. Petty Officer Johnston came up to us. He had this big welt across his face. Through everything, he came up with the radio. He said, 'I held onto it.' We tried to hail people on the radio, but it was dead."

At this time, John DeMello and the other two U.S. Coast Guardsmen learned the flashlight they had seen on the beach belonged to National Park Service Ranger Mark O'Neill, from the Mora ranger station.

"Mark O'Neill ended up running toward us. He took our belts that go around the [exposure coverall] and clipped all three together to make a sling for my arm. He said, 'Okay, we've gotta get back. Our only chance is running along the beach.'

"Will said, 'Look, that didn't work out for us the first time. Let's think of something else.'

"I said, 'I agree. I don't want to run along that anymore.'

"It ended up we climbed over logs to get off the beach. I felt my shoulder pop back into place. We walked back toward the truck. I remember jumping down into what is normally a small stream and ended up in waist-high water. The water had come up that far. I was shaking my head at the whole thing.

"I learned later that Petty Officer Mumford was going kinda of crazy because he hadn't heard from us. He had been trying to call us. The park ranger had got out before we did and told him we were coming. I came into the truck and I can remember Petty Officer Mumford's face.

"Petty Officer Mumford said, 'Dammit! Oh, man! Dammit!'

"Then he grabbed the back of my . . . [exposure coverall] and he pulled me up into the truck. I just laid on the seat. I was tired. I was still trying to put into place just what had happened.

"Petty Officer Mumford drove us back to the station. I walked into the station through the front doors. I looked into the comms room. There were Trevor and Zandra. I looked at Trev. He looked at me and said, 'Oh, man!' and turned around.

"Scooter came in with a big welt across his face. They called an ambulance for us. Fe walked in front of me. I was, like, look, I'm hurt. She just walked by me and she was mad. I didn't know why. I found out later that she had been on another beach party and they had performed CPR and she had gone through a lot.

"Will came up to me and put his hand on my leg and said, 'Look, John, they found someone on the beach.'

"I said, 'Who is it?'

"'Wingo.'

"Fe grabbed me. She thought I was going to punch someone, but I had one good arm and I wasn't going to do nothing. I started to cry a little bit. I think that was the only tears I shed until this day. I feel so bad I was the only one who has not cried for him."

I recall going onto the rec deck to sit with John until the ambulance arrived. When I passed Fe, she had tears streaking her face.

13 "We don't die, we save people"

While John DeMello fought on the beach, Captain Volk, at Port Angeles, recalled another helicopter crew from home and called the command center of the Thirteenth District in Seattle to request assistance from U.S. Coast Guard Air Station Astoria, Oregon. Volk attempted to send surface units to the scene. He requested the 110-foot patrol boat *Cuttyhunk* (WPB-1322), which happened to be at Neah Bay, to proceed to Quillayute River.

Chief Warrant Officer (CWO2) Robert "Bob" Coster, commanding officer of the Neah Bay station, later said, "It must have been around 1:30 in the morning when my telephone woke me up. I was about half awake when I identified myself. I heard, 'Bob, this is Phil Volk.' That woke me right up! Captain Volk asked me if I could get one of my motor lifeboats under way and proceed to Quillayute River. His voice was as emotional as I have ever heard it." Bob Coster went to the station, reviewed the conditions, and had to tell his Group Commander that "I could not send my boat, that weather and sea conditions were so bad my crew would be put into too much danger and, furthermore, could not get down there in time to help anyone." For those who understand the military hierarchy, for a chief warrant officer to tell a captain that he could not get his boat under way is a major decision.

Captain Volk next turned to Captain David Kunkel, commander of Group Astoria, Oregon. CWO2 Randy Lewis, commanding officer of the Grays Harbor, Washington, station, to the south of Quillayute River, found himself in a conference call with his Group Commander Captain Kunkel and Captain Volk. Captain Volk asked if CWO2 Lewis could get his 52-foot motor lifeboat, the *Invincible,* under way to Quillayute River. This was a logical decision on the part of Captain Volk: the surfmen, coxswains, and crews of the small-boat rescue stations in the Thirteenth District swear by the larger 52-footer. It can take an unbelievable amount of punishment. One reason for the larger motor lifeboat at Grays Harbor is the presence of a very bad bar. Lewis also had to say that conditions were so bad that he could not send his boat. I have heard from crew members at Grays Harbor who were willing to go, just as Coster's crew was ready to go, but both Coster and Lewis made the correct decision. It is difficult to explain to those who have not worked in maritime search and rescue the helpless frustration felt by the commanding officers and crews of both stations that night. CWO2 Randy Lewis, of course, had *two* captains he had to refuse. Much later, Phil Volk would tell me, "Both Bob Coster and Randy Lewis made the right decision that night." As Captain Volk pointed out, the commanding officer of the 110-foot patrol boat *Cuttyhunk* said "he'd stick his nose into the storm and see if he could get down to Quillayute River. As we know, he did and got the crap beat out of him."

Nineteen-year-old Fe Brantley, dressed in her exposure coverall, rode in a station vehicle to the First Beach area and then walked behind a police officer. "They had put on the lights from the ball field near First Beach, but it was still dark," Fe recalled. "The beam of light from the policeman's flashlight shone like the light sword in *Star Wars.*

"We came out of the tall weeds in the beach sand, and the first thing I saw was that Master Chief had pulled off his hat and was shaking his head. I think he was saying, 'I should have trained them more.'

"I didn't know what to say. I figured Master Chief and the XPO could take care of themselves. If it had been anyone below me, I would have tried to help them. I just didn't know what to say to Master Chief. I knew then that this shit was for real.

"I saw them doing CPR on someone. I thought it was one of the policemen they were working on, not one of our guys. Matthews had his . . . [exposure coverall] pulled half off. He was just standing there looking down with this look on his face. I figured something was wrong, because he was sobbing, crying.

"There was a big log between me and the person they were working on, so I went around the log. I looked down and all I saw was a . . . [exposure coverall] and feet out. I was tripping out because he didn't have any socks or shoes on. His rubber boots or felt boots, he didn't have them on. I thought at first it was Wingo. I ran down there and I was cryin'. I know at one time I was holdin' his head in my hands and I looked at his eyes and they were wide open. I was lookin' at his face, but not lookin' at it. I was looking at it out of the corner of my eye. His eyes were all cataracted over. I kept thinking that doesn't mean he's dead. That doesn't necessarily mean he's dead.

"I was trying to find his pulse. You could feel a little bit, but that's because of the CPR. I started praying.

"The tide was coming in, or something, because we got a little water on him. We picked him up. I remember thinking, 'Unconscious people you're supposed to cradle their head. You're supposed to do this, you're supposed to do that, just concentrate on CPR and get them breathing. Get them to wake up.'

"One of the guys was doing mouth-to-mouth. He had to stop because he was puking. Paul Lassila started.

"One of the guys who was doing compressions had to stop. I started to do compressions. I had to stop, because I couldn't deal with it. I wanted to be up there by his head talking to him. I kept thinking, 'We don't die, we save people.' I know I was just blabbering.

"When we picked him up, then I knew it wasn't Wingo. Wingo's solid. Then I looked at him. I said, 'Clint? Oh, no, it's Clint.' He was one of the babies of the station."

At this point in my interview, two years and four months after the loss of the crew of the ready lifeboat, Fe could no longer go on and started to cry. She took a few minutes to get herself composed and then hesitantly began again.

"I just knew Clint was not going to die because he was like Wingo, he was always happy. Clint was happy he got somewhere close to home. He can't be one of them. He can't die because he's so close to home."

Falicia again had to stop. She took a deep breath and began once again.

"I wasn't being real rational. I was rubbing his head and telling him, 'Oh, Clint, you're going to be okay.' I started crying really bad, so they told me to go back to the station. I went with the cook. By the time I left, I was sure he was going to be fine."

Charles A. Lindenmuth II, an EMT with the Forks ambulance corps, and his wife, Roseanne, after being awakened by advanced radio calls from the ambulance bringing in the two people from the *Gale Runner,* heard the police make a "dual tone alert"—a radio tone alert that is the signal for "an immediate emergency response (referred to as a code three) for crew number two." When Lindenmuth heard that crew two was responding, he and his wife relaxed a little. "The anxiety level was again raised when we heard a radio message from the other ambulance requesting that we bring plenty of extra blankets and hot packs as there may be as many as six more victims."[1]

Charles and Roseanne started toward LaPush. In the storm the "ambulance would shudder each time a strong crosswind hit it. However, the most unnerving thing was being physically slowed down when turns in the road caused us to head directly into the wind. The wind-driven rain was almost moving horizontally. . . . The rain was heavy enough to make me wish that there was another speed to the windshield wipers, and I silently promised to be a little less eager to jump behind the steering wheel the next time out."

The Lindenmuths were dispatched to the beach to pick up injured searchers on a National Park beach. "As we raced to the park beach, we turned up our multichannel scanner to try to hear anything about our potential patients. The next thing we heard from our scanner was the one phrase that makes already pumped-up hearts beat faster: 'CPR in progress.'"

After checking with the police dispatcher, Charles turned the ambulance around and headed toward where CPR was being performed on Miniken. The Lindenmuths pulled up to an area where there were "nearly 15 vehicles parked with their overhead lights on" and, grabbing their equipment, made their way to the scene across "a 12-foot-wide pile of driftwood."

A circle of people watching the efforts moved out of the way as the ambulance EMTs set up their equipment. "When the defibrillator leads were attached, the technician shouted over the stormy roar of the ocean for everyone to stand clear and then pushed the analyze button. The monitor screen showed asystole or flat line, indicating that no shock was needed. We immediately began to put our patient on a backboard so it would be easier to move from the beach. After looking at the path leading away from the beach, I decided to wait until we reached the ambulance to place an endotracheal (ET) tube in our patient. It would have been difficult, if not impossible, to maintain adequate placement of the tube while going over the big pile of driftwood.

"When we were ready to move, we passed the backboarded patient from one group of people to the next until we were clear of the woodpile where we had our gurney waiting. Within two minutes of leaving the beach location, we had our patient at the rear of the ambulance. . . . I placed the ET tube properly and then loaded the patient into the ambulance."

Charles realized they were going to have to perform CPR for at least 20 minutes before arriving at the hospital. He asked two of the Olympic Park rangers to ride along to help. The wisdom of this was soon realized. "Since the ambulance was not configured for a center-mount gurney, I was forced to lean across the lower torso of the patient to start the IV as one of the Park rangers helped me stay upright, tightly grasping [me by] my belt." While one of the rangers helped insert the ET tube, the other passed out needed supplies.

As the ambulance pushed through the storm, "the ET tube had shifted from its original position. There was so much sand on the patient, and all of us, that the tape normally used to hold the tube in place would not adhere to any surface. We rinsed off as much of the sand as we could, using bottled water carried on board the ambulance, and I re-intubated with a new and larger ET tube."

While the ambulance carrying Miniken sped toward Forks, Fe returned to the station. "[SN] Sara [Zurflueh] came up to me when I got back and shook me. She grabbed my hands and looked down. My hands were covered in blood. I didn't realize it at the time. She made me wash my hands.

"After washing my hands, I came back in and saw John and Scooter sitting there. I didn't see them when I came in; I guess I just blanked out there for a bit. I saw Scooter, who has these little drumstick legs, with a big knot on his leg. I said, 'What'n hell happened to y'all?' I was *so* mad at them. I kept calling them dummies because they didn't wear their SAR vests.

"Then John told me what happened when the wave hit him. I was irrational. I was just mad at the whole night. Scooter was hurting and I sat beside him.

"Will came in and told John it was Wingo on the beach and he was pretty sure he wasn't going to make it. John was crying.

"I said, 'What's wrong with you people? First of all, he is going to make it and second, it wasn't Wingo.' I was just mad. At everything and everybody."

The next ambulance arrived and took John and Scooter to the Forks hospital.

Meanwhile, Charles A. Lindenmuth and his ambulance crew with Miniken pulled up to the emergency room area of the small Forks hospital. "People came from almost every direction to help us move the stretcher into the ER." The crew continued CPR while the doctor contacted the "nearest level-one trauma center," Forks being a level four. "Except for infrequent pulse checks" the crew continued with CPR, recalled Lindenmuth, "until the youngest and strongest of us lasted for five minutes before requiring relief. We reached a point where we needed relief as our crew had been going for over four hours."

Then, sometime "during the midmorning the doctor called a halt to the resuscitation effort. The assembled people, in and around the ER, could see tears in other people's eyes, just like the tears running down their own cheeks. It was the first time, however, that we realized that four other patients were being treated by our little level-four hospital. The two injured rescuers from the other side of the river and the two people from the distressed sailboat had been undergoing treatment at the same time we had been working on our patient."

Later, John DeMello recalled, "We got to the hospital, and this is the horror part of the story for me. I went to x-ray and after that they put me in a room across the hall from someone. I hear them shocking him. I hear, 'Clear!' then the sound of the electricity. I put my head down and said, 'Please don't let that be someone from the station.' I'm thinking, 'Put me in another room away from here.'

"Scooter walked into the room and stood next to my bed. He said, 'Miniken is across the hall.' I heard two doctors talking while Scooter was with me. One said, 'We can't save him.' The other said, 'Yeah, it's over.'

"I remember looking into Scooter's eyes and he looked at me. We didn't say anything, we just hung our heads."

. . . 14 *Jayhawk* 6003

While Falicia Brantley fought back her tears on First Beach, 200 miles to the south Commander Michael A. Neussl, stationed at Group and Air Station Astoria, Oregon, and senior duty officer on this night, came awake to the SAR alarm and the announcement of a lost 44-foot motor lifeboat in the vicinity of LaPush, Washington. Commander Neussl donned his flying gear and hurried to the operations center, where the duty telecommunications specialist and Group duty officer briefed him on the missing motor lifeboat. Neussl then directed the launch of the ready HH-60J *Jayhawk,* number 6003, with maximum fuel of 5,600 pounds, FLIR (forward-looking infrared radar), night sun (a very strong spotlight), and an extra dewatering pump. Both Commander Neussl and his copilot, Lieutenant Michael T. Trimpert, wore night-vision goggles.

The *Jayhawk* left the ground at 1:50 A.M., and Commander Neussl estimated at least 50 minutes of flight time. En route Mike Neussl briefed his crew, consisting of Lieutenant Trimpert, Aviation Structural Mechanic (AM2) Richard J. Vanlandingham, and Aviation Survivalman (ASM2) James Q. Lyon, on what he knew. Additional information came in by radio during the flight. Neussl kept the 6003 at 500 feet on the flight northward over the ocean. The *Jayhawk* flew in bad weather: westerly winds at 40 to 50 knots, with heavy rain reducing visibility to 1 to 3 miles and the crew experiencing turbulence, especially when downwind of terrain. Closer to the search area, Commander Neussl again briefed his crew that he "intended to make all turns into the wind and that there were power lines extending out to James Island from LaPush."

At approximately 2:45 A.M., the Astoria helicopter crew arrived near James Island. Commander Neussl could hear the radio transmissions of the Port Angeles helicopter 6589 as they made their hoists of Marcia Infante and Kenneth Schlag from the *Gale Runner.* Neussl began his search in the immediate vicinity of James Island. The rescue swimmer operated the FLIR, which helps locate warmer objects in colder seas, while the aviation flight mechanic operated the night sun. Both Neussl and his copilot used night-vision goggles.

Group Port Angeles informed Commander Neussl that the second Port Angeles helicopter, number 6585, was en route to James Island and that the *Jayhawk*'s crew should search south of the island, while the Port Angeles helicopter would work the north. Mike Neussl set up two oval-shaped patterns, "modified to provide a safety margin" and turning "into the wind." On the second pattern "we spotted the sailing vessel *Gale Runner* awash on the beach and in the surf."

Into the third pattern and on the offshore leg, Neussl heard on the radio a transmission from "Quillayute River mobile 3 finding a motor lifeboat crewman on First Beach and that CPR was in progress. . . . We turned to return to the vicinity of First Beach to concentrate our search at this new datum. We did a low and slow hover search of the surf zone off the beach north towards the jetty, again picking up severe orographic turbulence when in the lee of James Island."

In the laconic manner of many pilots, Commander Neussl described a very close call. "We came close to the power lines strung across the entrance due to the strong wind and drift, causing us to perform a maximum power wave off to the left and away from the obstructions."

During the power wave off, Lieutenant Trimpert saw a flashing light through his night-vision goggles. Once leveled off, Neussl "maneuvered the aircraft sideways up and over the cliffs and trees on James Island . . . in order to investigate the light source which was at the head of the cove and up above waterline."

When the large helicopter drifted over the cove, the air crew spotted the missing motor lifeboat, the CG 44363. Neussl fought a drift of the helicopter to the right "by putting in left cyclic, to the point that we were 15 degrees left wing down, and we were still drifting right. I called for Lieutenant Trimpert to come on the controls and assist me in moving the aircraft back to the left before I lost sight of the survivor. . . . We required 20 degrees angle of bank to stop the right drift and move the aircraft slowly to the left, after which we repositioned the nose into the wind and aligned the aircraft with the major east-west axis of the cove, over its south-side cliffs." James Island's steep cliffs rise at least 183 feet into the air.

The air crew tried scanning the cliff for a survivor and attempted to see if anyone was still attached to the motor lifeboat. The air crew discussed whether they should try to put the rescue swimmer into the cove. Commander Neussl tried a low hover hoist and found severe turbulence below 220 feet. The pilot deemed it unsafe to put his rescue swimmer into the cove. Neussl, however, "felt it was extremely important to maintain visual contact with the suspected survivor since we did not know his condition and I didn't want him to give up the will to live if he saw the helicopter fly away, leaving him in the dark again and, possibly, leaving him for dead." Like many U.S. Coast Guard aviators who have accomplished something special, Commander Neussl's sentence is made to sound matter-of-fact.

What the air crew of *Jayhawk* 6002 accomplished could hardly be called "nothing special." In the early morning hours of 12 February, the strong winds were constantly buffeting and moving the helicopter, which has a gross weight of 21,246 pounds. Neussl and his crew fought the heavy machine, keeping it away from the dangerous cliff, so that Ben Wingo could have support.

During my time on active duty, my shipmates and I many times bemoaned the flight pay of the "airdales." In the early morning hours of 12 February 1997, I saw and heard U.S. Coast Guard helicopter crews more than earn their pay. One was recalled from home and flew search patterns in very bad weather, another made an unbelievable rescue, and the third, from another air station and Group, fought the dangerous turbulence off James Island so that an enlisted U.S. Coast Guardsman would know he had not been forgotten.

For over two hours, Commander Neussl and his air crew remained in a hover watching over Ben Wingo. During the long hours, the air crew spotted "two reflective pieces of material floating in the cove, possibly people in the water. The objects were being tossed about by the surf, unable to drift out of the cove." The terrain and turbulence made it impossible for the large helicopter to descend close enough to positively identify the objects, "but," said Commander Neussl, "we were fairly certain that we were looking at the other two crewmen floating in the surf."

. . . Part Three Aftermath

. . . 15 "Everyone's heart just sagged"

Prior to daylight on 12 February, Ben Wingo still clung to the side of James Island, and the bodies of David A. Bosley and Michael E. Schlimme had not yet been recovered. Outside people began to arrive at the Quillayute River station. Among the first were crew members from the Neah Bay station, Quillayute River's isolated neighboring small-boat rescue unit to the north. CWO2 Robert Coster, the commanding officer of the station, said at "least two-thirds of the ready crew" made the journey to the south. CWO2 Coster also sent his executive petty officer, BMC Philip E. Spurling, in charge of the detail. The Neah Bay people were to help with watchstanding, beach patrol, and, because the U.S. Coast Guard leadership wants mishap incident reports into headquarters rapidly, some were to help with the paperwork required. "I was proud of the way my crew volunteered without hesitation," said Bob Coster. Zandra Ballard remembers "a BM3 from Neah Bay putting his hand on my shoulder and just telling me that I was doing good and asked if I needed any help. At this point, I was thinking 'Why can't everyone just clear out of here?' Too many people. Several people tried relieving me. There was no way that I was leaving there until I knew everything was taken care of and in control."

My recollections of the period from the discovery of the bodies in the cove and Ben on the cliff until late in the morning are blurry due to lack of sleep. I recall coming onto the rec deck and seeing Commander Paul Langlois in his dry suit, looking exhausted from his ordeal. We said a few words, and Langlois said, "I was scared." It is the first time I have ever heard a U.S. Coast Guard pilot admit such an emotion. Paul, despite his weariness, set about helping the crew of the Quillayute River station.

Shortly thereafter, I talked to ASM1 Chuck Carter, the rescue swimmer on the Port Angeles helicopter that made the rescue off the *Gale Runner.* He looked at me and said, "Someone was watching over us." Later still, I talked with BMCM George LaForge. His face showed the strain of the long sleepless hours and the information that he had lost three crewmen.[1] After the passage of over three years, I can still see George's haggard face, close to losing control. George shook his head and said, "I should have trained them more. I should have trained them more."

Sometime during this period I recall walking back to get what seemed like my one hundredth cup of coffee and seeing one crew member start to break down; another was right there to comfort the person. Later, I saw a group sitting exhausted on the floor of the rec deck, trying to help each other.

Before daylight, some commercial radio stations were already broadcasting information about something happening at the isolated Quillayute River station. Zandra remembers around four o'clock in the morning "the mother of Petty Officer Lassila calling and wanting to know what was going on. She said she had heard about the accident on the news and wanted to know if Paul was involved in it. She didn't know anyone had been killed. I remember telling Petty Officer Cookingham about the call, and that is when they started to notify families. We were told that we could call our families just to let them know that we were okay and not to worry rather than being bombarded with family calls later on."

As daylight began filtering through the low overcast and driving mist, the second motor lifeboat, with BM1 Placido in charge, made its way across the bar and back to the safety of the station. MK2 Tom Byrd's experience on the motor lifeboat that night and next morning is typical. "I have never been so scared, or so sick, as that night. All I could think about was waiting for the next day and getting back across that bar and into the harbor. We were all bundled up in our bunny suits, and underneath that, thermal underwear, and we were still cold. We would go down into the forward cabin and try to get warm. It was pretty nice down there. I remember losing one of my gloves and my hand was so cold.

"The next morning, after it got light, that's when we actually saw how large the waves were. I just couldn't believe how big those waves were we had been out in all night.

"I remember how good everyone felt when we were finally able to get back across the bar. What happened really didn't hit us until we got back inside the harbor and turned into the boat slip and seen the 63 [the first motor lifeboat] wasn't there. I think that's when everyone's heart just sagged. We all were hoping it was there and then seen it was gone. We all knew something bad had happened. No one still knew what happened, until we got inside the building. That was hard. It's an experience I never will forget."

Near daylight, Chaplain Commander Gordon Scheible, the first of the Critical Incident Stress Management (CISM) team arrived at the station to help the crew through their grief. Chaplain Scheible, after driving approximately 150 miles to get to the station, said to me, "I didn't bother to stop at too many stoplights."

The CISM is something from the new Guard. In earlier days, people were expected to "suck it up," to use today's vernacular, and continue on with their work. Mostly, people went to a bar and got drunk. The CISM is usually made up of four people, generally a social worker, a chaplain, and two volunteers, who are chief warrant officers or higher. "The concept is to get people who are external to a unit, and sometimes external to a tragic scene, to be a listening ear, hopefully to have a member talk about his or her traumatic view of a situation."[2]

Fe Brantley recalled that, while everyone waited to find out who had survived from the first motor lifeboat, "I had to decide who to pray for. I had already prayed for Clint. Who did I pray for, Schlimme, Bosley, or Wingo? I liked Bosley, but I figured he was the oldest, so he could look down and forgive me for not praying for him to come back. I thought about Matt, and I loved Matt to death, I told him, you can forgive me. Wingo is 19 years old. I said he hasn't lived. Even the short life you've lived has been pretty good from what you've told me, and you can forgive me for praying for Wingo. So I concentrated all my praying on Wingo."

Daylight brought the Clallam County sheriff's high-angle rescue team to the station. The team would be helicoptered to James Island and then lowered onto the island. They would then deploy to get Wingo off the cliff.

Ben Wingo, in the rocks some 50 feet above the water, nursed a broken nose, a cut above his eye, and an injured leg. Later, he recalled that from the look of the coxswain's chair he thought Bosley was dead, but that Miniken and Schlimme had made it safely ashore.

Throughout the night, a helicopter hovered near Ben's location. Its spotlight flashed upon him and then into the waters below, back and forth. As the surrounding area started to show daylight, Ben stood up to relieve himself, and a helicopter came in close. "I was, like, oh now you come down here," Ben recalled.[3]

Ben spotted someone up on the edge of the cliff walking around and shining a flashlight. Ben started to yell and the rescuer asked him his name and rank. The rescuer told Wingo not to move. A second rescuer descended to Ben's location. He fitted Wingo with a harness, while the first person watched from above. The first man saw something on the rocks below and asked Wingo if it was a lifejacket. Ben looked down and saw Bosley lying face down "like he was comfortable" on some "jagged, jagged rocks." He told the rescuer that it was not a lifejacket.

The rescue team helped Wingo to the top of the cliff. The helicopter hovering above lowered the rescue basket. Ben climbed into it and was hoisted upward.

"It was so loud in there, and the mechanic pointed to the seat. I got out [of the basket] and he stored everything away. Then he looked over at me, grabbed my hand, shook it, and gave me the thumbs up.

"That's when it hit me what happened and I started crying."

The helicopter landed at the station's ball field. A waiting ambulance could have whisked Ben quickly away to the hospital in Forks, but Commander Langlois would later say, "I recall the helicopter saying they had Wingo and should they bring him to the airport and then the hospital? I made a quick decision and said, 'No, I want you to land here unless he is badly injured.' I wanted the station to have something to show there was some hope."

The ambulance backed up to the entrance of the station. The doors opened and someone in the group standing by the station's door said, "There's your shipmate." A group of people rushed the ambulance.

Fe Brantley recalled, "When Wingo stepped out of the ambulance . . . getting married, having my son, and that moment are the times I most remember in my life so far. He still had his SAR vest on and I jumped up on him. His shoulders are about an inch taller than my forehead and I got scratched on the face from the equipment. I think he is the only person strong enough to come back from that. I wanted him to know how I felt, but I didn't know what to say, and I'm never at a loss for words. I just hugged him."

Ben was brought into the station and onto the rec deck so more people could see him. Ben had the look that any veteran in the line would understand. There were cuts about his unsmiling face and his eyes were staring at something only he could see. When he came onto the rec deck, Ben would later say, "I knew it was serious when the XPO saw me and hugged me. The XPO doesn't hug *anyone*!"

Fe would recall, "When we got him inside, everyone was hitting him on the back. That was probably the best time during the whole night. All the officers ushered him into a room. We're standing outside the door. We didn't know what to do. When John [DeMello] found Ben was in the room, he knocked on the door and no one answered, so he opened the door and went in and hugged Ben." After the short reunion and a change of clothing, Wingo went back to the ambulance and to the hospital in Forks.

The last thing that I recall before I finally went to bed in the afternoon was the recovery of the bodies of Bosley and Schlimme from the cove. I slept only fitfully and finally gave it up as a bad try. When I walked out onto the rec deck things had changed dramatically. There were people everywhere.

Captain Volk arrived at the station with the Commander of the Thirteenth Coast Guard District, Rear Admiral J. David Spade. Captain Volk showed the same fatigue as everyone else who had been up all night, plus he had just made the long drive to Quillayute River with the admiral.

The media, in full frenzy, now entered the picture. A communications student at the University of Washington described the beginning of the barrage: News "helicopters were dispatched to the scene of the disaster. Satellite news trucks followed. . . . The reporters began to arrive . . . shortly after sunrise. . . . In the first few hours . . . [after the media learned of the incident] Coast Guard public affairs people fielded approximately 200 calls from both local and national media while attempting to prepare an official statement."[4]

The local media somehow gained access to the list of crewmen and "indicated that there were two survivors before the facts were in or the families contacted." Ben Wingo's girlfriend "was contacted by the media before it was known whether . . . [he] was alive or dead.

"Within hours of the deaths, the grandfather of . . . [Minikin] very reluctantly answered questions from the press, sometimes amidst tears. Subsequent media coverage of the memorial service focused largely on close-ups of those who were clearly grieving, as well as on the bruised and swollen face of . . . [Wingo]. What some members of the Coast

Guard 'family' considered private moments had become public statements."[5]

Captain Volk would later say, "I understand . . . [the media's] requirements and appreciate what they do. What I don't appreciate is their need, at any cost, to get a story. This was such a major event and because there were so many, they were everywhere. Even with the help of Public Affairs, they [the media] had needs and wants beyond that. My need was to feel my men were cared for and protected. I don't know if we are ever able to provide enough staff to keep the media satisfied."[6]

Commander Raymond J. Miller said, "I thought press handling of the incidents was mostly factual. But there were some exceptions. For instance, a local reporter came out and told me he realized I wouldn't want to be talking about the case but he wanted to know what the weather was like that night. When the piece showed up in the paper, he colored it. He indicated that I had said it was 'unexpected' or a 'freak' storm but I had never characterized it that way."[7]

The station put a guard on the government family housing area to prevent any reporters from trying to talk to people in the housing. Captain Volk said, "Master Chief LaForge's wife walked over to the post office and two reporters were waiting there, and they put a . . . [microphone] in her face and said, 'Tell us about your feelings.'"[8]

For the next several days after the incident Zandra recalled that the media "seemed to be very persistent in trying to get up . . . [to the housing area]. They sat across the road at the store and if a wife went to the store they would know it and just start asking tons of questions. One thing about the media, they were trying their hardest to talk to Ben. They somehow got the phone number to the pay phone [at the station] and started calling that and saying they were his sister or some relative. But through all of that they still didn't get to talk to him. I think that we screened the calls pretty well."[9]

Media bashing during times of disaster is easy, as there are always cases that seem inappropriate. Not all of the media in the Quillayute River incident deserve condemnation, however. Commander Miller recalled, "There was another time when the media could have jumped in and they chose not to. After the memorial service we put roses in the water as a symbol of the crewmen who died. While we were doing this there were a cameraman and two reporters there and they decided not to come up and talk to us. There were about 14 family members present, too. The press elected not to intrude."[10]

The pressure of the media coverage did not necessarily all come from outside of the U.S. Coast Guard. Captain Philip C. Volk recalls that on his long drive from Port Angeles to LaPush, with Admiral Spade in the car, "I took a badgering cell phone call from the captain in charge of headquarters public affairs. NBC had heard about the case. They had had some kind of a special 'in the can.' They'd had it on the shelf and hadn't done anything about it. Now, because of this case, they wanted to bring it out and do a special and show about LaPush. I would not give my permission to interview the crew. The captain and I had a discussion. He could not understand why I would not allow this to happen. I explained it the best I could." Recall, this is after Captain Volk had spent the night directing the search and recovery efforts.

"[The captain] hit me with 'there's a human interest story here.' The last part of that conversation was, 'I have a boss in the Thirteenth District. You call him and if he allows this to happen, it'll happen, but if you're asking me, based on the authority I have as a group commander, the answer is no, and have at it, pal!' Click! Captain Terry Sinclair, the district chief of staff, supported me."

A few petty officers at Quillayute River recalled that after the crowds arrived some officers took the training records and the station's log book without notifying the unit. It was not until after Ben Wingo applied for school in the aviation field that anyone noticed his physical examination information was missing from his medical record.

Admiral Spade gave a press conference near the front steps of the station. Zandra Ballard would remember that Admiral Spade was "visibly upset . . . about the whole accident." At this conference the admiral could only give the media what little was known at the time.

Sometime during the day, a boat crew from the Grays Harbor Station, to the south of Quillayute River, brought one of their motor lifeboats to the station and remained there to assist the station for a period of time. When BM1 Dan Smock, a surfman, and his relief crew, came into the station carrying their SAR bags, the camera people were tripping over themselves trying to film them.

My last two recollections of 12 February came in the evening. In spite of all the rumors of the strained relations between the tribe and the station, the first people outside of the U. S. Coast Guard to do something for the unit were some women from the tribe who came onto the mess deck carrying a great deal of food. Later still, Master Chief LaForge, BM2 Brent Cookingham, a tribal policeman, and I were standing on the station's front steps talking while a large television relay van settled in just outside the station. "Goddamn vultures," said Brent. "I wish we could get rid of them."

"We can't do that," said George.

"Yeah, but *we* can!" said the policeman. "We've just been waiting for you to say something and we were going to close off the road so they couldn't get down here."

I went to bed shortly thereafter, and on Thursday, 13 February, with the investigators and even more people arriving, I left for home. I knew the last thing LaForge and Placido needed was an additional person on the station.

I could not, however, divorce myself from the unit. I continued to think about Wingo, LaForge, Placido, Cookingham, Meterko, Ballard, DeMello, Brantley, Lassila, and all the others. On Sunday, 16 February, I drove to the station for a short visit to see how everyone was holding up.

Not surprisingly, George had not slept much in the intervening days. He informed me that the official memorial service would be held on Wednesday 19 February and that there was also to be a private station memorial service, because Christina Schlimme would be taking Matthew home to Whitewater, Missouri, for burial before Wednesday and this was a chance for the crew to say goodbye to the man who tried to make the crew laugh. George invited me to attend, saying, "You were here that night too."

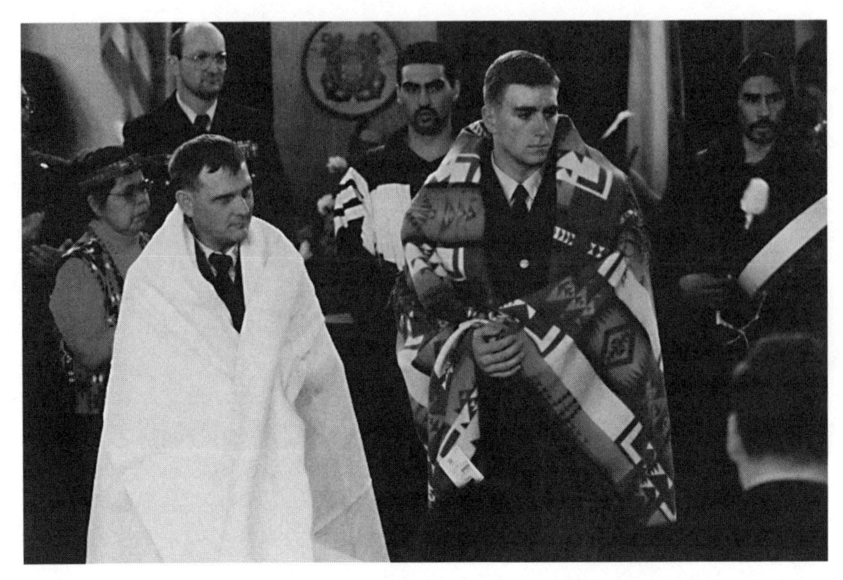

Fig. 15.1. BMCM George A. LaForge (*left*) and SA Benjamin F. Wingo receive blankets from the Quileute Tribe at the memorial ceremony for the loss of the three crewmen. The blankets are given to ensure they will be safe in life. Photo courtesy U.S. Coast Guard.

As families and friends began arriving for the scheduled 1:00 P.M. memorial service, the Grays Harbor crew took the communications watch and served as the ready boat crew. With everyone settling in, what happened next would not be believed if it appeared in a Hollywood production: The SAR alarm blared out of rec deck's public address system. "Person in the water off Second Beach." The look of fear on the faces of the wives is something I will remember for a long time. Some covered their ears. I could hear the Grays Harbor crew racing for their boat. There was just the slightest hesitation on the rec deck by some people and then Jon Placido's voice boomed out: "All right people! Let's go!" Don Meterko remembered that it was a close race between the two crews as to who would be out first, but the Grays Harbor crew had the jump and were followed very closely by the Quillayute River crew.

The Gray's Harbor crew pulled the person out of the water, backed up by the Quillayute River boat. There is nothing like a "save" to buoy up any crew. It was interesting to view the faces of the Grays Harbor crew when they returned to the station. Most were trying to suppress a grin—you could see them working at trying to play the serious-faced, macho, just-another-day-at-the-station attitude. The salts were better at it than the newer crewmen. The little swagger in their walk, however, gave away their feelings.

After everyone had returned and things settled down, the station's memorial began. The proceedings were very emotional. I did not look forward to the official memorial on Wednesday.

The official service proved a surprise because of the great number of people who attended. The Quileute tribe had offered the use of their community center (fig. 15.1). Secretary of Transportation Rodney Slater, Admiral Robert E. Kramek, the Commandant of the U.S. Coast Guard, in addition to Senator Patty Murray, Rear Admiral J. David Spade, and a host of others would be the senior people at the ceremonies. There were representatives of all the small-boat stations in the Pacific Northwest and other units of the service. The Canadian Coast Guard sent a delegation, and the law enforcement departments of the area also sent representatives.

The memorial service put yet another level of stress on those at the unit. Fe Brantley recalled, "The Commandant and secretary of transportation wanted to speak to us before the ceremonies. The Master Chief said, 'The Commandant is going to say a few words to you.' He came in so quick that we didn't stand up for him and we got chewed out because of that. I wasn't even thinking about rank or nothin' right then.

"They only talked for a few minutes, so I am all mad at the Commandant." Fe also echoed what other crew members would tell me: "When the commandant gave the final salute, I thought why didn't he salute them, they had their hats and medals on a table, he could have saluted them, but he saluted the television camera. I couldn't stand that."

Fe also recalled, "[A]t the memorial service, if you saw someone crying, they [the media] were right there. They took pictures. One reporter asked for my name. I said, 'You can read my name tag.' When he asked for my first name, I declined. He asked two more times."[11]

Zandra would remember, "I thought Admiral Spade was very sincere in everything that he did. He seemed to have a lot of compassion and feelings towards all of us."

One of the more emotional moments that many did not know occurred when the sole survivor of the 52-foot motor lifeboat in the *Mermaid* case in 1961 introduced himself to Wingo. Gordon E. Huggins said to Ben, "I know what you are going through."

Ben Wingo would recall of the public ceremony that, at first, he could not cry (figs. 15.2 and 15.3). "I was, like, 'What's wrong with me?'" Then the chaplain read a passage written by a U.S. Revenue Cutter Service officer in 1885 about another crew of maritime rescuers from the U.S. Life-Saving Service: "These poor, plain men, dwellers upon the lonely

Fig. 15.2. SA Benjamin F. Wingo, the only survivor of CG 44363. His face still shows some of the cuts from his ordeal. Photo courtesy U.S. Coast Guard.

Fig. 15.3. The crew of CG 44393 after receiving the U.S. Coast Guard Achievement Medal for their work on the second boat on 12 February 1997. Master Chief George A. LaForge is at far left. The crew (*left to right*) BM1 Jonathan A. Placido, MK2 Thomas L. Byrd, BM3 Marcus M. Martin, and SN John A. Stoudenmire III. The memorial to the three lost crewmen is in the background. Photo courtesy Thomas L. Byrd.

shores . . . took their lives in their hands, and, at the most imminent risk, crossed the most tumultuous sea . . . and all for what? That others might live to see home and friends."[12]

Wingo finally cried. Bosley's widow, Sandi, hugged him. "Don't feel bad one bit," she whispered. "You made it, honey. You're such a good boy."

16 Investigations

There were three investigations into the incident at Quillayute River. One, called the safety investigation (headed by Captain Carmond Fitzgerald, Commander, Group, Detroit, Michigan), is classified. Captain Fitzgerald explained that the safety investigation was classified so that people would feel free to talk; thus, there was a better chance to learn what happened and to improve any safety defects. The second, known as the administrative investigation, is open, and most of my comments on the investigation come from that report. The third, also classified, was known as the Commandant's Vessel Safety Board (CVSB).

The administrative investigation was performed by a board made up of three officers. Commander James M. Hasselbalch headed the board, and Lieutenant Polly P. Bartz and Chief Warrant Officer Two (CWO2) Terry E. Smock made up the members. CWO2 Smock, commanding officer of the Coos Bay, Oregon, station, was the only one of the three officers who had small-boat experience. The board was selected and ordered by the Commander of the Thirteenth Coast Guard District, Rear Admiral J. David Spade. The board would gather evidence and interview people. After finishing this task, Commander Hasselbalch would send his report to Admiral Spade. The report would have a preliminary statement, findings of fact, opinions, and recommendations.

Once Admiral Spade received the investigative report, he would comment upon the report, agree or disagree with the findings of fact, opinions, and recommendations, and then forward it to Admiral Robert E. Kramek, the Commandant of the U.S. Coast Guard at the time, for the final, and official, comments.

It seems obvious that in the Quillayute River incident the administrative investigation would look very closely into at least five areas: the weather, condition of the boat, the boat crew, the crew's training, and the command's policies and procedures. Information on the weather will be discussed later. The board quickly established from station records that the 44-foot motor lifeboat CG 44363 had no known defects that could have caused the accident.

Commander Hasselbalch's investigation found some fault with the record keeping of the station. Training records showed that Bosley had not been recertified for coxswain for his six-month period beginning 1 January 1997. The records also showed that he had not completed a nighttime familiarization. Schlimme had not been recertified as a boat engineer. BM2 Donald Miterko, the training petty officer for the station, said he was "behind on his paperwork," that both Bosley and Schlimme had completed all their requirements before the 31 December 1996 deadline, but the entries had not yet been made.

No one seemed to think Master Chief LaForge's training program was at fault. At least one of the investigators called it "a near model boat crew training program."[1]

In the naval services, the person in charge of any type of craft, or ship, is held responsible in an accident, even if the person was not the direct cause of the accident. It is not unusual, then, that the investigation would focus sharply on BM2 David A. Bosley. Placed under extreme scrutiny, the 36-year-old petty officer presented a troubling picture. In fact, the probing of the boatswain's mate's record drew attention away from other matters that should have been equally troubling but that received scant comment from the media and official sources.

Before discussing David A. Bosley, it is necessary to understand two traits of those serving in the U.S. Coast Guard. First, there is a very strong mistrust of investigations; second, military people will speak in a code if they must say something about a shipmate and do not want to give the appearance of lying or saying anything bad about the person. This is a code that must be deciphered in order to understand what was being said. The depositions of the crew members to the board of investigation provide very few negative comments about Bosley. The worst: "he was a cowboy" and "Bosley's adrenaline gets flowing and he yells a lot." Most of the crew members of the Quillayute River station did try not to say anything negative about Bosley, yet put out hints for those who were willing to listen. For instance, "Bosley returned to the station screaming and yelling, 'Where's my crew?'" This statement could indicate a man who may be too excitable to stop and think clearly.

More than two years after the deaths, some former crew members still felt reluctant to speak negatively of Bosley. One person said, "People will tell you they liked Bosley, but they really didn't." More than one person said, "A lot of people didn't like Bosley, but I did." A summary of their comments indicates that he could be personable, but that he lacked "people skills." Many stated he could be overbearing, and one person said, "he did not react well to stress."

The most damning evidence against Bosley came from his own service record. In a move condemned by many U.S. Coast Guard enlisted people, headquarters released otherwise restricted information from that record. Bosley's record showed that on 8 March 1995, he had his coxswain and officer of the day qualifications rescinded by the previous officer-in-charge, BMCS Daniel E. Shipman. Apparently, Senior Chief Shipman—in what the U.S. Coast Guard terms a Page 7 (Administrative Remarks) entry—recorded that Bosley had been "involved in a personal campaign of deceit, intimidation and fear" and that this had taken on more importance to him than his "duties as Officer of the Day, Lifeboat coxswain, and as a petty officer." Furthermore, the "campaign" was "motivated by your quest for power over others no matter the cost." Moreover, Bosley, according to Senior Chief Shipman, had "disregarded . . . specific orders and regulations by openly criticizing or denouncing their need or purpose, failing to properly execute them, and substituted them with your own as you have seen to fit your needs."[2]

Senior Chief Shipman's remarks continued by pointing out that Bosley had "used intimidation and fear to cover up your mistakes and bolster your ego while underway on the unit's boats." Shipman also indicated that he been made aware of "numerous situations" that "clearly show a lack of judgment and disregard for standard practices which places the crew and boat at extreme risk."

Shipman closed with, "I can not and will not tolerate this behavior." He concluded that Bosley was "entrusted with command of a vessel" and that he bore "the responsibility of the lives, both physically and mentally, of your crew and are responsible for their safety, health and well being. You have forsaken all this."

Sixteen days later, Senior Chief Shipman counseled Bosley for departing on liberty and leaving the ready boat "not ready for sea." According to the entry, 11 of the 16 bolts that hold the forward handrail down were missing. "This created an extremely unsafe condition and shows a lack of judgment." Shipman also indicated that at the time the coast "was experiencing heavy swell conditions. If the MLB [motor lifeboat] was dispatched to sea, the likelihood of severe damage and personnel injury could [have] occurred due to the absence of these bolts." On 18 July, Bosley was again counseled, this time on his failure to meet the recertification requirements for boat crewman.

Shipman would later admit that Bosley did make an effort to improve and eventually regained his qualifications. Yet it was not long thereafter that additional adverse comments were put into Bosley's service record. On 26 April 1996, Shipman entered comments on Bosley for "standing a poor watch and not keeping the SDO informed of changes which require notification of the chain of command and changes in unit readiness." The comments are almost prophetic: "On the morning of 26 Apr 96, your first light bar check was at 0545. You reported to the Communications Watchstander that the bar was 4–6 ft. At 0600 the XPO drove in and told you to go check the bar again because it is bigger than was posted on the board. The 0600 bar report was 10–12 [feet], occasional 14 [foot] breaks. The SDO should have been informed due to the requirement of a surfman on board during breaking bar conditions. You did not inform the SDO at all and he didn't leave his house until 0640. This put the unit and personnel at risk by not having a qualified surfman aboard." Senior Chief Shipman also indicated that this had not been the first time Bosley had neglected his duties in this way; Bosley had also not reported a barge dragging down river.

On the same page, Senior Chief Shipman also noted that on 12 April Bosley had taken the RHI 205539, a fast, small outboard-powered boat, to the mouth of the Quillayute River to do a bar report. According to the entry, the bar was breaking at 12–14 feet and the river current was 8 knots. The officer-in-charge remarked that taking the boat out in such conditions was contradictory "to sound seamanship practices, outside your personal ability and qualifications, and extremely unsafe. By doing this you put yourself and another shipmate at risk." Furthermore, Bosley had not informed the SDO of his actions. Senior Chief Shipman maintained that when confronted with these and other charges, Bosley would become "evasive" and always had an "excuse."

The Page 7 entries continue on 18 October 1996, now with Master Chief George A. LaForge as officer-in-charge. LaForge's entry indicated that Bosley had been placed on report for improperly handling classified material—always a serious incident—on 23 September 1996. The charges had been dismissed with a warning. LaForge felt that Bosley focused his attention in too many directions, that classified material needed his total attention, and that Bosley needed "to ensure that you are not distracted by less important duties."

A law school education is not necessary for understanding the importance of these entries. Taken in total, they alone would seem enough to stop any further examination into the case by the investigating board. It is perhaps understandable why the news media quickly closed the story once these entries became public. But Bosley is so important to the events of 12 February that it is necessary to explore further his service career.

Chief Warrant Officer Two (CWO2) F. Scott Clendenin commands the Yaquina Bay station, located at Newport, about halfway down the Oregon coast. If any person personifies the spirit of the small-boat rescue stations, it would be Scotty Clendenin. He is among the top, if not the top, decorated U.S. Coast Guardsman serving at small-boat stations. While a chief petty officer, Scotty received the Coast Guard Medal, the highest award for valor an enlisted man at a small-boat station can receive in peacetime. Scotty won the medal for taking a 30-foot surf rescue boat across a breaking bar, rescuing five people from a capsized fishing charter boat in high seas and bringing the survivors safely across a bar that had at least 20-foot breaking seas. It took him three tries before he could get across. Then, he went out again to help in searching for further survivors. Yet Scotty admitted that the rescue he is most proud of is the saving of a small child earlier in his career at Tillamook Bay, Oregon. He has a fierce dedication to search and rescue and has chosen to devote most of his time and energy to the stations. He is one of the few people who actually lives up to his reputation. One rarely sees such determination and love for a profession.

In 1999, CWO2 F. Scott Clendenin granted an interview on David A. Bosley. At the time, Scotty had 25 years in the U.S. Coast Guard, with 18 of those years running motor lifeboats. Most importantly, he had been a chief petty officer and the executive petty officer at Yaquina Bay when David A. Bosley served under him, working on his coxswain and surfman qualifications.

"Bosley was hell on wheels," according to Clendenin. "He was hard on people. He did not have a perception of what people thought when he said something. He would say, 'Pick up that line' and that is what he heard. What other people heard was: 'You pick up that line, or I'm going to kick your ass!'

CWO2 Clendenin maintained that it was not Bosley's boat-handling abilities that brought him problems, as he was very good at handling a boat, but "sometimes Bosley would piss off people no end." Clendenin felt it important to recognize that when Bosley "got in trouble, it was not for the way the case went, but the way he treated his people." He also indicated that when Bosley served under him, Clendenin "spent most of my time with him working on the people thing. In a nutshell, he was a very competent boat driver, if you kept up on him. When his imagination was alive, you started chasing after him. He saw things different than other people saw them. His priorities were a lot different than others. He had a lot of energy, but a leader has to control that energy to make sure it is directed in the right area. When that energy was kept in the right direction, Bosley was a highly productive person."

Clendenin described Dave Bosley as "very intimidating. Question him, and he didn't see anything wrong." He told Bosley, "Dave, you're the high ball bat material. One day they're going to take a ball bat and beat the shit out of you." Bosley responded to Clendenin, "You can't talk to me like that."

"That would shock [Bosley] and we would go into a nice normal time," recalled Clendenin. "Then he would be in my office again and I'd say, 'You're on the ball bat list again, huh?' Dave was a high maintenance person." Scotty defined "high maintenance" as you had to watch a person constantly.

Even after these observations and descriptions of Bosley, Clendenin said, "If his name came up for my crew, I would ask for him by name. I probably had more time under way with Dave Bosley than anyone else in the Coast Guard. I know he could navigate."

BMCS Daniel E. Shipman, the officer-in-charge at Quillayute River prior to BMCM George LaForge, had 17 years in the U.S. Coast Guard when he took command of the station in 1993. "I fired my first XPO at the station," Shipman told me. The interview of Dan Shipman occurred in 1999, after he had been promoted to master chief petty officer and commanded the Tillamook Bay, Oregon, station. "I took a lot of heat for firing [my XPO], which builds into a lot of reasons why I made the decisions I did later on," Dan said.

"I took a massive amount of heat from the Group commander and XO [for my actions]. [This was before the arrival of Captain Volk and Commander Langlois at the Group.] At the same time I fired the XPO, there was an incident in Port Angeles where an enlisted man was murdered by his wife. The captain and XO were so distraught over the incident they couldn't function. It took 10 to 14 days for someone to come out and do an investigation. I never got a copy of the investigation into my relieving the XPO, nor never got to see a copy of it. My feeling at the time was the officer sent to investigate the incident was sent out to investigate me, rather than the person fired. I got a letter of administrative reprimand put in my file for my inappropriate behavior for firing this guy because I undermined his authority. What had happened was I had to run the station while waiting for the investigating officer. So, I had a third class helping me recheck all the files the XPO had messed up. That was how I undermined his authority. This about a guy who would sit out in the smoking area and tell seamen, 'Hey, the senior chief can't do that to you, he can't treat you like that.'

"I took that real hard. I was looking out for my crew, and this kid goes to the Group about how mean I am and they believed everything he said until the end, when he started to get in trouble at the Group. Trying to fire someone else after that with that regime in place was a no win situation for me.[3] I had to weigh that into my thoughts concerning Bosley. I would not be here in the Coast Guard today if I had went that route with him; they would have relieved me.

"Bos was a unique individual. When he first came aboard from Newport, I got a call from the detailer saying, 'I am sending you this guy, he's married with no kids, and is just about surfman qualified.'"

Dan went on to say that Bosley at first seemed to be a "very personable individual. I found out later that he was like a Dr. Jekyll and Mr. Hyde when it came to talking to me or the XPO and the rest of the crew. Placido saw through this and we started documenting his behavior."

Shipman related that when Bosley first arrived at the unit a crew member had reported that he had run the motor lifeboat aground.

Bosley denied the charge. Shipman told me he "tried to believe the guy; after all, he had just got to Quillayute River. We found out when we pulled the boat for the yards he had hit the keel hard."

"Bosley had a tendency to sugar coat the truth a bit," said Shipman. What finally made things worse for Bosley was the way "Bosley was victimizing people. Bosley had a power thing. He would say, 'I don't give a fuck what those people say, you're going to do it my way.' He was doing that on the boats. He was doing things not right on the boats. He pulled a couple of bonehead maneuvers with the boats." Shipman said that after seven months at Quillayute River he had Bosley's qualifications pulled.

Dan Shipman said that Bosley wanted to be a surfman. After seven months at the unit, Shipman realized Bosley was not going to be a surfman. Shipman told Bosley he would "call him" on everything he did wrong.

Bosley started to improve. "People at the station knew if he did what he did before, it would come back on him. He started to realize this and I saw some changes," Shipman said. "We were without a boat driver, we took that out of hide. It was quite difficult for us. Bosley was showing improvement, and I talked with Placido about it for a couple of hours. We decided that Bosley was doing better, let's give him the incentive that he has the coxswain quals back. I don't think I gave his OOD qual back."

Later, Bosley had other problems. Shipman and Bosley discussed his latest problems. Dan told Bosley, "'Three strikes and you are out. I will pull your quals forever.' When I left Quillayute River, he had two strikes."

Shipman explained all this to Master Chief LaForge when he relived him. "I told him here's what's going on. Keep an eye on Bosley and you gotta ride real close on him and keep him honest, because he can easily waver.

"I always felt there was a place for Dave in the Coast Guard, but it wasn't in charge of a boat crew," Shipman said. Dan felt Bosley would have been good on a ship under supervision. "Bosley could be very personable. It wasn't that he was stupid. He just was not cut out to be in charge of a boat crew." Shipman pointed out that Bosley was very good with his hands and woodworking. He felt Bosley should change his rate to damage controlman, and that would have suited him very well.

"In hindsight, I could have just gave up on him and hung him up," said Shipman. "I look back and think, I should have done that. At the time, I looked at it as here's an individual who has some worth, was doing what he wasn't supposed to be doing, but could be a benefit to the service in other areas. He wasn't just a total misfit.

"Then you look at the personnel assignment policy. I would have gotten a third class off a ship with no qualifications. I would have been set further back by firing him than keeping him around and trying to work with him. The detailers would not support transferring people out who could not cut the mustard."

BM1 Jonathan Placido said that Bosley bragged, "I was trained by Scotty Clendenin, you don't have anything to teach me." Jon also said that Bosley informed him he did not want to be a surfman, as he did not want the extra duty and responsibility. Another important point that Jon revealed: Bosley "met the Commandant's minimum requirements for coxswain, not master chief's, or my minimum requirements," so it would be very difficult to fire him as "he was still doing his job." Indeed, many of the responses from headquarters about stations note, "they meet the minimum requirements." Minimum requirements appear to be an acceptable standard for small-boat stations, according to headquarters.

Chris Koech, a former crewman, said of Bosley, "I didn't hate him, but the SOB pissed me off. But you have to remember that I worked closely with him almost the entire time I was stationed there. I don't care who you are, if you work that close to someone for too long, you're going to rub each other wrong eventually.

"I also must be honest that Dave was a hell of a good coxswain. I think that's about the biggest thing I found redeeming about him. We ran a couple of cases together, and I have to say that when he was behind the wheel I had no doubts that the case would be run well. Dave was prior service, marines, and, as such, he was old school. No matter what BS went on, when push came to shove, duty always came first. When he was behind the helm, he was behind the helm. Dave had pulled some crap before, but despite that you knew you could count on him when it was a case." In short, David Bosley is not easy to classify. A detailed look at the various scenarios of what could have happened that night suggests that Bosley may not deserve all the blame for the deaths.

Commander Hasselbalch's investigative board worked against a time deadline. The letter appointing him to head the board also ordered that he would file his report by 10 March, 25 days after the incident. There has been no official explanation as to why the investigation had to be completed within 25 days. As it turned out, the board needed until 19 March to file the report.

The narrative of the completed report consisted of 25 pages, with 70 enclosures, filling a very thick three-inch binder. The enclosures included everything from the sworn statements of witnesses to the autopsy protocols of Bosley, Schlimme, and Miniken. Perhaps it was the haste in which the investigation needed to be completed, but a close examination of the report raises questions. Maybe once the board saw the entries in Bosley's service record, they felt there was little else that needed to be accomplished.

Commander Hasselbalch's opinion section of the report to Admiral Spade amounted to 20 items. The board found that Bosley "saw the 1640 [4:40 P.M.] update to the National Weather Service 'forecast' at 1740 [5:40 P.M.] on 11 February 1997. BM2 Bosley had ample opportunity to inform both the Senior Duty Officer . . . and BMCM LaForge about the exact nature of the forecast weather conditions. If this information had been provided to either the SDO or BMCM LaForge, a surfman would have been onboard the Station that evening."

The board admitted that Bosley thought the *Gale Runner* was close to the Quillayute River bar, which "helps to explain the sense of urgency that BM2 Bosley felt about getting MLB 44363 underway."

The investigative board charged that Schlimme's remark of "Let's get the fuck out of here," had been "made in an effort to get BM2 Bosley to turn back and not attempt a bar crossing." The board also stated that the navigational lights on the CG 44363 were not turned on; that shortly "after crossing the bar but prior to the boat's first roll, the crew of MLB 44363 was disoriented and lost situational awareness."

The board's findings about Bosley's role in the incident: "BM2 Bosley did not have enough rough weather bar crossings at night in a 44' MLB to prepare him for the conditions that MLB 44363 encountered . . . and should not have attempted to cross the bar. . . . If a surfman had been on MLB 44363 that morning, the casualty would not have occurred." One surfman's opinion, however, is that "no one *ever* has enough rough bar crossings at night to be prepared." The surfman also said, "If you want to know how it is to do a night rough bar crossing, try getting in your car on a very dark night, turn off your lights, and drive down an unlighted freeway, then imagine the freeway moving."

Finally, the board stated "the proximate cause of this casualty was the coxswain's failure to safely navigate MLB 44363, causing the boat to capsize and founder in the surf conditions near James Island."

Commander Hasselbalch's board recommended that "all small-boat crew members attend Team Coordination Training (TCT) . . . [and the] Commandant publish clear, reasonable policy guidance regarding the wearing of safety belts and helmets." Other recommendations concerned personnel matters, aids to navigation, and paperwork concerning certification.

Admiral Spade finished his portion of the investigation and submitted it to the commandant on 28 March, 18 days after Commander Hasselbalch's report. In general, Spade agreed with the findings of fact with two small modifications, and added five additional findings dealing with personnel, the 44-footer, and whether BMCM LaForge had ordered Bosley to cross the bar.

In the opinions section of the investigation, Admiral Spade offered more about Bosley. He agreed that "BM2 Bosley's failure to inform his superiors about the deteriorating weather conditions the evening of the incident is consistent with his previously documented failure to report such matters as required." Furthermore, Bosley "was not aware of CG-44363's proximity of James Island when it impacted . . . [a] rock."

Commander Hasselbalch had reported that if a surfman had been aboard the motor lifeboat, the accident would not have occurred. Admiral Spade, however, remarked, "Generally speaking, surfmen have the advantage of greater training and would thus have a higher probability of success in conducting this mission safely; a surfman did in fact cross the bar safely afterwards. However, surfmen are not invincible, and they can make human errors in judgment and boat handling. That said, I am of the opinion that, had a surfman been at the wheel of CG-44363 that morning, the casualty *probably* would not have occurred." [Emphasis placed by Admiral Spade.]

In addition to the investigative board's comments about Bosley's failure to safely navigate the CG 44363, Admiral Spade's most detailed charge virtually indicted BM2 Bosley: "BM2 Bosley failed to exercise proper judgment expected of a Coast Guard coxswain and OOD by:

—Failing to keep track of his own currency training requirements and not self-reporting discrepancies to his command. . . . The element not documented in his recertification was at least one of three night-time operation/familiarization sorties in the 44-foot MLB. This shortcoming directly relates to BM2 Bosley's qualification to operate as a coxswain on the night of the incident.

—Failing to adequately inform his superiors about the deteriorating weather and sea conditions.

—Failing to properly brief, coordinate, and prepare his boat crew.

—Failing to require his crew to wear helmets.

—Failing to navigate safely.

—Failing to assess the rough bar conditions correctly and/or exercising inappropriate judgment by crossing the bar. He should have known the bar was too rough for a coxswain without surfman qualifications to attempt."

Admiral Spade also charged that Bosley "was no doubt motivated by the noblest of intentions—to save others in peril upon an unforgiving sea. Unfortunately, while rushing out with the best of intentions, he failed to recognize his own limitations regarding his qualifications and experience handling the boat in these conditions. Therefore, I find that the proximate cause of this accident was BM2 Bosley's failure to conform to the standard of care of a reasonably prudent Coast Guard coxswain."

In the matter of whether Bosley should have been retained at the unit prior to the accident, Admiral Spade wrote, "it was BMCM LaForge's responsibility to constantly assess whether BM2 Bosley's judgment, training and skill merited retention of coxswain qualification. Other than the one missing training requirement—and in view of the *many positive* statements by fellow crew members about BM2 Bosley's performance—I will not second guess BMCM LaForge's long, successful career as a small-boat operator; his daily interaction with his crew, his local knowledge, and his position as OINC made him the best person in the Coast Guard to make this tough call." [Emphasis placed by Admiral Spade.]

Before discussing the release of reports after Admiral Kramek's decisions in the following chapters, it is necessary to explain two phrases that will appear often: risk management and team coordination training. These phrases play an important role in how headquarters arrived at some of their decisions when releasing the final results of the investigation into the Quillayute River incident. To understand the strong emphasis on the "new" aspect of small-boat operations, one must turn back to the days of the U.S. Life-Saving Service.

Public affairs officers in the "old" U.S. Coast Guard were enamored with a phrase coined by a keeper in the old U.S. Life-Saving Service: "You have to go out, but you don't have to come back." Usually, public affairs officers would somehow manage to include the phrase when writing about saving lives from the small-boat stations. Under the "new" U.S. Coast Guard, the phrase is used pejoratively and "risk assessment" is in.

U.S. Coast Guard headquarters apparently maintains that the small-boat community does not take proper precautions in evaluating whether or not to send a boat out, thus the need for risk management, which has crews assigning numerical values to six factors. The assignment is usually made when crew members come on duty, and the values are 0 through 10, with 10 being the highest risk. The six values are added, and the GAR (green, amber, red) Evaluation Scale is consulted. The six factors are supervision, planning, crew selection, crew fitness, environment, and event/environmental complexity. The GAR Evaluation Scale is measured: 0–23 as green, or low risk; 23–44 as amber, or caution; and 44–60 as red, or high risk.

Team coordination training—an even newer phrase—is combined with risk assessment and apparently means that everyone should have the right to say something is wrong, even though the coxswain is still in charge of the boat.

In order to show that risk assessment and team coordination training are the correct procedures, headquarters apparently belittles the old U.S. Life-Saving Service keepers and the U.S. Coast Guard chiefs who ran the small-boat stations; they do so by making them seem either irresponsible or careless with the lives of their crews. When trying to revise a bureaucracy, policymakers often deride the old practices in order to make the new look more attractive. In reality, the old keepers and chiefs never willingly sent their men out into a gale knowing they would never come back, if for no other reason than they went out with their men. Previously, they used their years of experience and knowledge to judge whether or not to go out. They did not need a formula; it was second nature. Furthermore, the service realized the need to go out. In 1899, after a keeper failed to go out to a wreck, the U.S. Life-Saving Service published in their regulations for action at wrecks that "In attempting a rescue the keeper will select either the boat, breeches buoy, or life car, as in his judgment is best suited to effectively cope with the existing conditions. If the device first selected fails after such trial as satisfies him that no further attempt with it is feasible, he will resort to one of the others, and if that fails, then to the remaining one, and he will not desist from his efforts until by actual trial the impossibility of effecting a rescue is demonstrated. *The statement of the keeper that he did not try to use the boat because the sea or surf was too heavy will not be accepted unless attempts to launch it were actually made and failed,* or unless the conformation of the coast—as bluffs, precipitous banks, etc.—is such as to unquestionably preclude the use of a boat." [Emphasis added.] This regulation continued into the 1930s.[4]

Another phrase in the U.S. Coast Guard's lexicon appeared during the investigation into the Quillayute River incident. Vice Admiral James M. Loy, at the time of the investigation the chief of staff of the U.S. Coast Guard, released a "lessons learned" brief to all units of the service prior to the announcement of the administrative investigation. Admiral Loy admitted that Bosley "used an ineffective decision strategy by moralizing the urgency of the search and rescue case without adequately weighing the risk."

"Moralizing," as defined by headquarters, is a judgment that does "not consider safety risks, and are best embodied in the statement, 'You have to go out.'" Note the reappearance of the old phrase. Admiral Loy intimated that Bosley was focusing on the possibility of people in imminent danger, yet failed to consider safety.

There is nothing wrong with risk management and team coordination, as it is equally important to know when *not* to go out as it is to know when to go. Despite what some at headquarters may think, however, risk management has always been practiced by keepers, officers-in-charge, and commanding officers; it was just not codified.

. . . 17 Questions

The crew of the Quillayute River station suffered four recognizable periods of high stress from this incident. The first, naturally enough, was the loss of their three shipmates; the memorial service provided the next event; the third period, on 21 April 1997, would also prove to be a harbinger of the fourth, the release of the official investigation. On 21 April 1997, before the official notification of the administrative investigation, the chief of staff of the U.S. Coast Guard, Vice Admiral James M. Loy, distributed internally "Chief of Staff's Final Decision Letter of a Class 'A' Mishap: Loss of Station Quillayute River MLB 44363 and Subsequent Death of Three Coast Guardsmen on 12 February 1997." Admiral Loy's letter, apparently based upon the results of the safety investigation, appeared a little more than two months after the deaths.

Admiral Loy wrote that the "mishap was caused by a series of human errors." Foremost in these "human errors," according to the admiral, "was a combined disregard of risk assessment by elements of the Search and Response system (including the group, station and coxswain) during mission planning, crew selection, and environmental evaluations. This is evidenced by the lack of adequate briefings to evaluate the nature of the distress, the on-scene weather, and the capabilities of the potential response platform prior to launching the response asset."

Second, "the coxswain lost situational awareness. He failed to ascertain the weather as that requiring a surfman, and failed to identify/confirm adequate navigational references during the transit out of the river. Indications were that the coxswain never knew the position of the boat in relation to the surrounding hazards at any time following the river transit. After turning to the west, he drove the MLB too close to James Island."

Third, Admiral Loy charged that Bosley "exhibited poor judgement by not recognizing the situation was beyond his capabilities and experience, and yet proceeded to transit the bar." Furthermore, Bosley "used an ineffective decision strategy by moralizing the urgency of the search and rescue case without adequately weighing the risk."

According to Admiral Loy, the "lessons learned" from this incident proved that the "importance of conducting proper risk assessment and using risk management skills cannot be overstated. Utilizing these tools at the command and control level (district, group and station), as well as at the coxswain/crew level, is paramount to ensuring effective analysis and safe, successful missions." Loy also stressed that everyone "must understand and respect the limitations of their personal abilities." Interestingly, most of the actions that Admiral Loy suggested and felt would prevent a recurrence appeared in the release of the official results of the investigation almost two months later.[1] Furthermore, after an exchange of letters with Captain Volk, Admiral Loy modified his statement to read, "There was a combined disregard of risk management by elements of the Search and Rescue Response system during mission planning, crew selection, and environmental evaluations. This is evidenced by the lack of adequate briefings to evaluate the nature of the distress, the on-scene weather, and the capabilities of the potential response platform prior to launching the response asset." In this amendment, Admiral Loy does not proclaim that the group or station made a mistake, but he does admit "elements of the Search and Rescue Response system" did not practice risk management.

As this long report was an internal publication, the civilian world did not know that Bosley had been made the scapegoat, or that the station and group had been censured. This criticism produced additional stress for the station. The next event, the releasing of the report of the administrative investigation to the public, provided a catalyst for additional stress for the Quillayute River station personnel.

On Wednesday, 18 June 1997, slightly more than four months after the deaths, Rear Admiral J. David Spade called a news conference in Seattle to release the official results of the administrative investigation. Admiral Spade read from the commandant's report, which ran slightly over three pages. Admiral Kramek, as had Admiral Spade, noted that as "we assess the circumstances of that stormy night, let it be clear that nothing we say or conclude in the aftermath will ever diminish the bravery and dedication of Petty Officer David Bosley, Petty Officer Matthew Schlimme, and Seaman Clinton Miniken." However, he continued, "Unfortunately, the weather conditions were extremely dangerous, beyond the capability of the assigned crew. As we can best determine, the motor lifeboat was not safely navigated, resulting in the loss of life for three of the four crewmembers assigned."

Admiral Kramek agreed there was insufficient information to form an opinion about what Schlimme meant when he said, "Let's get the fuck out of here." Furthermore, Kramek modified both Commander Hasselbalch's and Admiral Spade's comments on whether the accident would have occurred if a surfman had been aboard. The official statement became, "I am of the opinion that, had a surfman been at the wheel of CG-44363 that morning, the casualty may not have occurred." In short, Commander Hasselbalch, Admiral Spade, and the commandant, Admiral Kramek, had all concluded that BM2 David A. Bosley caused the accident.

Admiral Kramek ordered a "Coast Guard wide safety standdown for all small boat stations and [to] ensure full compliance with all of the directives" he issued. Among these directives, the commandant ordered "all small boat crew members attend Team Coordination Training (TCT)"; clear, reasonable policy guidance be published regarding the wearing of safety belts, helmets and protective clothing; people who were E-6 surfmen no longer had to fulfill the sea duty requirements to make chief petty officer; there would be a study of the aids to navigation at Quillayute River; the bar lights would be restored;[2] there would be an evaluation of the handle for the forward main deck watertight hatch; U.S. Coast Guard Headquarters, Office of Operations, would "clearly define the terms 'rough bar conditions,' 'extreme offshore sea conditions,' and 'excessive river current conditions,' and the terms and their definitions and the restrictions and conditions associated with them would be fully distributed throughout all small boat stations"; there needed to be a reevaluation of the boat crew helmet design; and "that headquarters program, rating, work force, and assignment managers work with Coast Guard District Thirteen to establish a clear policy on the number of surfmen required at all Coast Guard District Thirteen stations based on individual unit circumstances."

Given access to Bosley's service record entries—combined with the comments from the investigation and the commandant of the U.S. Coast Guard—the news media quickly concluded that no matter how often officers stated that BM2 David A. Bosley was a hero, the blame for the deaths rested squarely upon his shoulders. The news media can, perhaps, be forgiven for not digging deeper into the story. The carefully choreographed release of the accident report, along with Bosley's service record entries, plus the final report, would seem to most people a cut-and-dried story. This is evident in a Seattle newspaper's announcement: "Coast Guard blames crew in LaPush fatalities." At least one newspaper praised U.S. Coast Guard headquarters for the "candor of Coast Guard investigators."

Fireman Apprentice Zandra Ballard, who had volunteered to take the communications watch during the morning of the deaths, was on her regular communications watch at the Quillayute River station the day the results of the investigation were released. "I got a phone call from a local who was yelling at me," recalled Zandra. "She had me in tears over what she was saying. She basically said that we were the ones that put fault on Bosley and blamed the entire crew for going against him. She said that if Sandi killed herself that it would be all of our faults. She told me that it is nice that someone who is no longer here and can't defend themselves, instead gets backstabbed and bashed by the service that they so proudly were a member of."

Other crew members told of being stopped on the streets of the town of Forks and having similar comments made to them. Tom Byrd, for example, said, "A lot of people in Forks blamed us about the investigation and it made you feel bad. It was hard dealing with that."

Zandra Ballard's remark about no one to protect the station from the calls prompted an inquiry to the Commandant of the U.S. Coast Guard about lack of support from the district's public affairs office. The question, in part, stated: "a crewmember at the Quillayute River Station who was on communications watch spent approximately four hours responding to questions from the community that stated: 'why did you [the crew] sell out Bosley?' . . . Surely, the district and headquarters must have known there was intense media interest in the case. Further, the district office is not that far from Quillayute River. Why was there no one in place at the station to assist when the investigation was released to the media?"

The response illustrates headquarters' lack of understanding of the close relationship between the stations and the communities they serve. Headquarters' Office of Operations provided the following answer to the above question: "The Public Affairs staff at the Thirteenth Coast Guard District was continuously available to provide support to Station Quillayute River personnel, if requested. No one was placed at Station Quillayute River to assist with answering media inquiries, in part because the station did not request assistance. In addition, because the station did not convene the investigation, they were not the appropriate Coast Guard unit to either receive or respond to media inquiries on the investigation. These inquiries should have been directed to the Thirteenth Coast Guard Public Affairs staff."[3]

This statement ignores the fact that stations have small crews. In the case of Quillayute River, the unit lost three people, had two injured, and was still trying to maintain their operational readiness. Yet, according to headquarters, they now were supposed to be public affairs specialists as well. This expectation stretches credulity, especially when the Thirteenth Coast Guard District has people trained in this field who could have assisted the station.

Apparently headquarters did not understand the question; the question was about the community. Most watchstanders know that *media* questions get passed on to higher authority, yet there was no one to take haranguing calls. While a public affairs person could not have lessened the comments station members received on the streets, they might have prevented the communications watchstander from suffering additional trauma. Instead, the crew shouldered the additional stress. When a captain at the Thirteenth Coast Guard District finally realized that civilians from the surrounding communities were calling into the station to harangue the crew, his staff responded: "Oh, Forks is antigovernment" and "Quillayute River is a unique community." The comments demonstrate a limited understanding of how the stations interact with their communities. Officers-in-charge and commanding officers have often commented, "Headquarters does not realize we have to live in these communities."

Dan Miniken, father of Seaman Clinton Miniken, felt "his son should have had additional training before being assigned to a dangerous job at a Coast Guard lifeboat station. . . . Everybody at the helm of a Coast Guard boat should be trained for any weather condition." The Coast Guard spokeswoman from the Thirteenth Coast Guard District, Chief Petty Officer Carolyn Cihelka, said, "To say everyone should be a surfman, maybe in a perfect world, but that's not going to happen."[4]

Bosley's wife, Sandi, decried, "I don't know anybody that's infallible. I don't know anybody who doesn't make mistakes. Of course, it's entirely possible that David made mistakes." She believed, however, that the deaths were "an act of God" and that the U.S. Coast Guard's report on the accident "unfairly" accused her husband, "while failing to note his awards for bravery. It's much easier to kick a guy that's gone and can't defend himself than some guy who is still here and can stand up [for himself.]"

Sandi Bosley also rationalized that Bosley's "superiors could have ordered her husband and crew back to LaPush harbor after they had encountered heavy weather. There was a time when his boat could have been recalled. The decision wasn't made. The person who could have made that decision must have thought it wasn't life threatening. There might have been times when David could have used more caution," she admitted, "but that's a big difference from saying he had a reckless disregard for his life or other people's lives." The charges in Bosley's service record "probably stemmed from personality conflicts with his supervisor at the time and his impatience with some of his crewmen. . . . He was not a sheep. He was very bright. He was very intelligent. He was never tolerant with people who weren't bright or intelligent."[5]

It was probably inevitable that many from within and outside the U.S. Coast Guard would be displeased with the official conclusions of the Quillayute River investigation. Part of the unhappiness results from what was published, while another part of the dissatisfaction comes from how those that serve at the small-boat stations feel they are treated. Still another aspect of the discontent is intertwined with the way many feel search and rescue from small-boat stations seems to be rapidly changing within the U.S. Coast Guard.

The published investigation raises several questions, so that in June 1999 I requested an interview with the Commandant of the U.S. Coast Guard. Admiral Loy, who had then been promoted to commandant, had the option of responding to questions by mail instead of by oral interview. After a long delay, the admiral's press assistant, Commander John Philbin, replied on 28 July that he had "been tied up with other things" and hoped to review the request "within the week." It was not until 16 September, however, and after more prompting, that Commander Philbin responded with "Admiral Loy's schedule is really busy. . . . If you send me the questions, I'll be happy to turn them around quickly with an opportunity for follow up attributed to Admiral Loy."

On 23 November 1999, Commander Philbin again responded: "I have not forgotten this. I'm having some internal problem with the program manager. I am pressing forward and will get your responses as soon as they get them to me." On 10 January 2000, nearly *four months* after I was promised a quick reply, headquarters responded: "I regret the delay that our slow response has caused you; however, I assure you that the Commandant does care to respond. I know that you are looking for personal responses from Admiral Loy but it is important to place the responses you seek in an organizational context. As I recall, many of the questions you posed addressed issues during the time Admiral Loy was the Chief of Staff—not the Commandant. Additionally, your questions crossed organizational program responsibilities and require coordinated input from several offices. I have been advised that the response is very near completion and we will continue to make every effort to expedite its delivery."[6]

On 7 February 2000, some *five months* after the promise of a quick reply, I consulted the Master Chief Petty Officer of the U.S. Coast Guard, the senior enlisted person in the service and advisor to the commandant, Vincent W. Patton III, and seven days later Commander Philbin responded by e-mail. "Attached is Admiral Loy's response. A signed copy is in the mail. Rear Admiral Riutta's response should be in the mail soon and I will e-mail you a copy as soon as it is signed. . . . I wish you well as you complete your manuscript and hope that you will consider carefully the responses that are provided by the Assistant Commandant for Operations."[7]

Sources contend that Loy did not feel it appropriate that he respond to an incident that happened while he was not commandant. The admiral, however, was chief of staff at the time and signed off on the important "lessons learned" document published by the service. Admiral Loy also later used the Quillayute River deaths in one of his speeches.

One of the first questions concerning the investigative report: Why did the investigation have to be completed so quickly? Many inside and outside of the small-boat community described a June 1997 Coast Guard helicopter disaster in which an HH-65A from the Humboldt Bay, California, air station crashed into the sea approximately 60 nautical miles west of Cape Mendicino, California, killing all four U.S. Coast Guardsmen aboard. That investigation lasted more than a year before the final results were known.[8]

When queried about the length of the Quillayute River investigation, the Assistant Commandant for Operations, Rear Admiral Ernest R. Riutta, replied, "The time taken to complete an investigation varies with the complexity of the issue being investigated. We do not believe that the time taken to complete [the] investigation was inappropriate. No valid conclusion can be drawn by comparing the time taken to complete this investigation with the time taken to complete any other investigation."[9]

Another interesting question concerns the lack of comments in the investigation from the Group Port Angeles command and control structure. Each of the three officers from the group had important roles in the case: Captain Volk was SAR mission coordinator during the case, while Commanders Langlois and Miller not only made an amazing rescue, but later worked at the station itself. Yet none of these senior officers was questioned. When asked about this, Admiral Riutta responded: "The investigating officer is no longer in the Coast Guard. Therefore, we were unable to interview him."[10]

The investigative board had interviewed witnesses either in person in a motel room in Forks or by telephone. One of the stranger facts about the investigation is that Lieutenant Kenneth Schlag, the owner of the *Gale Runner*, had only one telephone interview, and the testimony amounted to one printed page. Marcia Infante, the passenger aboard the *Gale Runner*, received two telephone interviews, resulting in three and a half pages. Four and a half pages seems very little testimony from two of the more important witnesses in this case. The Coast Guard witnesses from the station were asked detailed questions on their role, while Schlag and Infante received, at best, superficial inquiries.

Weather obviously played an important part in this incident. Yet references to environmental conditions come from the Quillayute River station weather log entries and a copy of the National Weather Service's forecasts. There was nothing in the report concerning reflective or refractive waves.

When questioned about the role of weather and waves, Admiral Riutta responded, "Although the weather was a significant factor in this mishap, the investigating officer apparently made the judgment that the analysis of specific weather conditions—such as wave characteristics—were less relevant to the investigation than the decision making process of the coxswain who faced those weather conditions."[11]

More than two years after the incident, an article in the *Seattle Times* noted, "As a result of the tragedy, the Coast Guard instituted broad changes. It stepped up replacement of the self-righting boats with faster, safer, longer ones. It replaced the old surfbelts hooks with new self-locking models and made training and safety procedures mandatory, emphasizing that surfbelts and helmets must always be worn in heavy seas."[12] In addition, the Coast Guard waived the requirement for qualified surfmen to have sea duty in order to be promoted to chief petty officer. Of all the items enumerated for improvement at his press conference, Admiral Spade felt this was the most important item.

When questioned about the *Seattle Times* article, Admiral Riutta stated: "The Coast Guard instituted specific changes to some policies, procedures and processes based on recommendations made by the Coast Guard's Vessel Safety Board, convened to review the investigation." The admiral continued to insist that the wearing of surfbelts and helmets was "reemphasized or clarified" during the safety standdown. Admiral Riutta wished to emphasize that no changes had been made to the delivery dates of the 47-foot motor lifeboat schedule. He also indicated that the surfbelts were "redesigned" with a double-locking quick-release safety buckle. Moreover, the definitions of "heavy weather" and "surf" were further clarified, "for the purposes of risk assessment decision making." When "breaking surf greater than 8 feet" is encountered, or expected to be encountered, a motor lifeboat must "have a certified surfman aboard."[13]

Many surfmen and crewmen found the new hooks extremely difficult to use, especially when their hands were cold and wet, which is common in the heavy surf of the Pacific Northwest. Only three years after headquarters had ordered the stations to change to the new belts, they withdrew the belts from service, saying that "the hook itself has been proven difficult to use in a cold environment with gloved hands and plastic portions of the hook easily break. Internal latch gate lock components rapidly wear out enabling the latch gate to be opened without depressing the latch gate release."[14] One senior petty officer remarked: "Each motor lifeboat in the Coast Guard has six belts. They made each unit buy the belts and then did not listen to what the people in the field said. Then they remove the belts. And they wonder why we are broke." Another portion of the protective gear mentioned in the published report, the helmets, have not been changed, even though the autopsy indicated that the death of each of the crewmen was caused by trauma to the head.

The statements about protective equipment insinuated that there were no training programs or safety programs, and that neither surfbelts nor helmets were required. Master Chief LaForge's statement on the evening before the incident, that safety was the most important thing in surf drills, disputes this belief. The same can be said for the use of surfbelts and helmets. It is difficult to explain why Bosley did not have his crew in helmets. Several officers-in-charge have commented that the irony in the lack of safety gear is that it happened at a station commanded by the man who had been in charge of standardization in the U.S. Coast Guard; LaForge had a reputation for writing up any station that did not put on helmets and surfbelts. Just because Bosley's crew had not worn all the safety devices available does not indicate that no policy existed. The coast guard comments about safety procedures and the mandatory wearing of the belts appear to be a method of reconfirming policy.

The passage in the *Seattle Times* article about the replacement of the 44-foot motor lifeboats "with faster, safer, longer ones" is also an interesting issue. The U.S. Coast Guard has, for the first time since 1915, paid for a private firm to build a new motor lifeboat to replace the 44-footer. Since the first of these boats officially went into service, in June 1997 at Cape Disappointment, Washington, they have been the subject of serious controversy. Some of the criticism came from the small-boat community because of their reluctance to change—there had also been derogatory comments about the 44-footer when it replaced the 36-footer. One briefer in U.S. Coast Guard headquarters emphatically insisted that "there is nothing wrong with the 47-footer." Reports from the small-boat stations insist otherwise.[15]

The coast guard has gone to great lengths to describe the virtues of the 47-footer. On 19 June 1999, for example, Rear Admiral Paul M. Blaney, Commander of the Thirteenth Coast Guard District, reported that "The 47's are replacing the out-dated, 30-year-old, 44-foot boat that has served well, but is too slow, too cumbersome, and less safe than the upgraded 47."[16] If the 44-footer were "less safe," why did the service keep it in operation for so long? Moreover, why do other agencies within the United States, as well as some Central American countries, want to purchase the surplus boats? According to the investigation into the Quillayute River incident, it is apparent that the 44363 did not fail the crew. Wingo recalled shutting down the engines of the boat. Furthermore, the boat had not sunk, even though it had been subjected to stresses and conditions more strenuous than it had been designed to take. Very few boats can withstand crashing onto the rocks. The Quillayute River station was not scheduled to receive new 47-footers until May and September 2001, making it the last motor lifeboat station in the Thirteenth Coast Guard district to receive them.[17]

Admiral Riutta's response about the 47-foot motor lifeboats: "The 47-foot MLB design includes navigation, speed and buoyancy enhancements when compared to the 44-foot MLB. This in no way suggests that the 44-foot MLB is unsafe in any way, but merely reflects advances in technology and design over the last 30 years. The 44-foot MLB is being replaced due to the platform's age—the boats are approaching the end of their serviceable life."

The questioning next turned to BM2 David A. Bosley. Evidence given to the admiral indicated that Bosley no longer aspired to being a surfman. Admiral Riutta was asked why Bosley was not transferred. Additionally, the admiral had been questioned about BMCM Shipman's statement: "The detailers would not support transferring people out who could not cut the mustard."

The admiral's official response revealed that "We have no information that BM2 Bosley did not want to be a surfman, nor are there any records to indicate he desired an early transfer from Station Quillayute River. All information that we have indicates he was performing satisfactorily as a 44' MLB coxswain, and it appears that he was scheduled to complete a normal tour of duty. He was recommended for sea duty upon completion of his tour, which is the normal practice to facilitate meeting advancement criteria for his rate." Moreover, "the policies and practices at that time were designed to keep the stations manned to their authorized levels to the best extent possible, and that included a requirement for qualified coxswains who were not surfmen. Further, there was no requirement for all coxswains to be participating in the surfman syllabus. Those who either did not desire to be a surfman or who were not able to satisfactorily complete any part or all of the syllabus, were removed from the program and served as regular coxswains so long as they were judged as competent to do so. In March 1999 we changed our assignment policies to allow transfer of those who do not show satisfactory progress in the surfman syllabus. We did this as part of our overall goal to increase the yield of the training process."

Part of my questioning included the comment by the former officer-in-charge of the Quillayute River station that "The detailers would not support transferring people out who could not cut the mustard." Admiral Riutta did not respond completely to the question, but did remark, "If a station OIC or commanding officer ever judges a coxswain to be unqualified to serve in that capacity there are very clear guidelines and policies describing how to proceed, and the assignment system always supports the final determination. Unqualified personnel are reassigned."[18]

Unfortunately, it took the assignment system five months to replace Bosley after his death. BMCM Shipman also revealed that if he had replaced Bosley, he probably would have gotten an untrained boatswain's mate and would have been in worse shape. In the meanwhile, Shipman would have been operating with even fewer people.

My questioning then turned to why, after over two years, the station still had only three surfmen instead of four. Admiral Riutta replied, "Our surfman training program currently cannot produce enough personnel for us to fill all the billets to the desired levels and there are not enough already qualified personnel at the proper pay grades to meet all of our needs. We have been and continue to analyze the system in an attempt to remedy this situation and we are making some progress. Until that is accomplished, we have developed policies to manage the risks to the maximum extent possible with the resources that we have. For example, we recently modified our assignment policies to retain surfmen in surfmen positions longer which has helped to stabilize this force as an interim measure until the training system can be improved and yield enough to meet our needs. We have also made changes to the training program to slightly increase its capacity, but we still have more work to do. I do not have an estimate of when the system will be what we need it to be, but I do know that we are working diligently to solve the problem."[19] According to many surfmen in the Thirteenth District, for years they have been advocating longer tours for surfmen but their suggestions have gone unheeded.

Admiral Riutta commented specifically as to why there were still only three surfmen at Quillayute River: "The manning requirements . . . represent the desired optimal state, but we are still able to meet minimum mission requirements by modifying watch rotations, training schedules, etc., in order to maintain at least minimum mission readiness at all stations. If a station OIC, commanding officer, or anyone else in the chain of command determines that a station is marginal or below minimum conditions of readiness, we work together to either solve the problem or have the unit stand down until readiness is reestablished to an acceptable level. This is not a good situation, because it increases risk, it takes a toll on our people, and it reduces our ability to train additional surfmen."[20]

The best translation of this official lingo comes from retired Captain Philip C. Volk, former commander of Group Port Angeles:

> One of the real tragedies at small boat stations is that, by default, the officer-in-charge and executive petty officer become regular watchstanders. When the weather is bad, everyone is on port and starboard. [One day on and one day off.] Sometimes LaForge and Placido were on "port and report." Figure out what the watch rotation is when you only have three; or when someone is on temporary additional duty, or sick. Nowhere else in the Coast Guard, except patrol boats when under way, do we require the CO and XO to be a regular part of the watch list!
>
> Certainly the officer-in-charge and executive petty officer should be "qualified in type" as we say in aviation, and they should stand some duty. Even I stood SDO on a rare occasion, but usually with the Safety Officer, or Engineer Officer riding in the right seat to keep me out of trouble and never when the weather was too crappy. But why in God's name we require the OIC and XPO to be regular watchstanders in addition to their other duties is beyond me. I remember weeks at a time when LaForge was the second boat surfman because his other two surfmen, the XPO and the BM2, were standing port and starboard as first boat surfman, in addition to leading, managing, and training. Amazing! And it gets worse.[21]

Admiral Riutta's response to the surfman question cites "Minimum mission requirements," which means he is able to "meet minimum mission readiness at all stations." Many officers-in-charge and commanding officers complain that "if we trained to commandant's standards, more people would die. We have to train to higher standards." It is ironic that in an organization that touts itself as the "premier maritime service," the leadership will accept minimum standards for the units that do the most search and rescue in the service.

Before his retirement, Master Chief LaForge retorted, "What bothers me is how people are using my dead crewmen for their own agenda." Indeed, other people at the station also complained that the incident produced statements such as: "Because of Quillayute River." The best illustration of LaForge's comment is included in an article in the service's official magazine. In the June 1999 issue of *Coast Guard,* Admiral Loy wrote, "Tragedies, like the Quillayute River case in Oregon [*sic*], in which three Coast Guardsmen died, have demonstrated the urgent need for a recapitalization of our communications and response system and urge us to improve watchstander training staffing."[22]

When John DeMello—the fireman apprentice who heard the initial call from the *Gale Runner* at Quillayute River—read this passage he was visibly upset. Nothing in the published investigation disparages the communications watchstanding during the *Gale Runner* case. Nothing had been said about communications in the published investigation. The audio levels of the transmissions were difficult to hear, not surprisingly, but the weather had been very bad and, most importantly, the radio antenna of the *Gale Runner* had been carried away with the mast. Even with the most up-to-date communications equipment, anything received from the sailboat would have been bad. Moreover, the tapes of the transmissions, made in Port Angeles, are remarkably discernable, despite Marcia Infante's use of a hand-held communications device. Admiral Loy's rationale for mentioning communications and communications watchstanding in the context of the *Gale Runner* results from the 29 December 1997 case of the sailing vessel *Morning Dew.*

The sailboat *Morning Dew,* en route from Myrtle Beach, South Carolina, to Jacksonville, Florida, had one adult and three teenaged boys aboard. At 2:17 A.M. on the morning of 29 December, a rapid and broken radio transmission was received on the distress frequency at U.S. Coast Guard Group Charleston, South Carolina; it said, "MAYDAY! U.S. Coast Guard! Come In!" The lone communications watchstander could not discern the call and felt the call actually said "U.S. Coast Guard, U.S. Coast Guard." The watchstander made three unsuccessful attempts to contact the caller. After the attempts, the watchstander did not notify the Group duty officer.[23]

At 6:28 A.M., more than four hours later, the same watchstander answered a call from a Charleston pilot dispatcher "informing him that a crew member on an inbound ship thought he heard cries for help as his ship transited the mouth of Charleston Harbor." The pilot boat accompanying the inbound ship was directed to search in the area and then call back. The watchstander made no connection between this call and the earlier one. He informed the Group duty officer of the pilot boat searching for the cries of help, and then a new communications watch began.

At 7:15 A.M. Group Charleston learned that "the pilot boat had searched the area of the reported screams without results." The Group duty officer was still unaware of the radio call of the previous night and did nothing further. "At this point, the Coast Guard had still not opened a case, and the Group Commander was still not aware of either event."

Four hours later, at 11:15 A.M., the "GDO received a call from the local police reporting that the bodies of two young males had been spotted and recovered from the surf from the vicinity of Sullivan's Island [near the entrance to the harbor]." All four people aboard the *Morning Dew* had perished.

Admiral Loy uses the *Morning Dew* to illustrate the need for better communications and communications technology. Yet, these are two radically different cases and there is really no excuse for the commandant to interchange the two when trying to make a point.

When questioned directly about communications, Admiral Riutta responded, "Although the investigation does not refer to communications, or communications training, the fact remains that during this incident, the parent command [Quillayute River] did not know where the CG 43363 was located, except when radio contact was initiated—and position information provided—by the coxswain. Automated asset tracking is a part of our National Distress Response System Modernization Project (NDRSMP). Had asset tracking been in place, it is possible that the parent command could have provided preventive guidance to the coxswain regarding this process." This comment caused one surfman to exclaim angrily, "This is just playing *Star Wars* bullshit!"

Admiral Riutta continued, "In the excerpt from the Commandant's speech that you quoted, he stated that *watchstander* training and staffing needed improvement. [Admiral Riutta's emphasis.] From this, you inferred communications training and staffing. In the Quillayute River incident, there was a failure of the coxswain to exercise proper risk management, and insufficient team coordination. In addition, we cannot rule out the possibility that more experienced watchstanders—at various levels in the search and rescue system—might have encouraged the coxswain to conduct a more thorough risk assessment of the situation. They may also have helped to avoid the boat launch altogether by warning the *Gale Runner* of the weather and hazards associated with attempting to enter port."[24]

Admiral Loy and Admiral Riutta imply that watchstanders at other levels were not trained enough to stop Bosley, or even the *Gale Runner* from trying to enter Quillayute River. Remember that the administrative investigation board took only one page of testimony from Schlag, the owner of the *Gale Runner.* Without a detailed statement from Schlag and Infante for reference, it appears, according to the radio transcripts, that the people aboard the sailboat were exhausted from fighting the weather and were intent on entering the harbor. At the time of Infante's first call inquiring about the bar, there were no restrictions. John DeMello informed the Group communications watchstander, Telecommunications Specialist Marshall, of the severity of the weather. Marshall told DeMello to relay this information on to the *Gale Runner.* Further communications revealed that the boat was coming in and was at the bar. And despite Admiral Riutta's denial, watchstanders had tried to stop Bosley. The Group and station both radioed Bosley to stop. Furthermore, Master Chief LaForge radioed Bosley to go to the bar and look the situation over. We will never know why Bosley continued, just as we will never know whether Bosley received and/or understood the instructions.

Because the administrative investigation failed to ask questions of the command and control structure of Group Port Angeles, the record does not include the evaluation made by Commander Raymond J. Miller when he was first notified of the call from the *Gale Runner.* Even so,

Bosley still got under way. All of this suggests that Admiral Loy and Riutta mistakenly thought that more highly trained watchstanders could have prevented the deaths.

The risk assessment made by the two chief warrant officers, Bob Coster of Neah Bay and Randy Lewis of Grays Harbor, when their Group Commanders asked them if they could get their boats under way, has been downplayed. Although their crews were willing to make the attempt, both commanding officers judged the risk too high.

Another of my questions asked why, if communications needed improvement, the U.S. Coast Guard has not used station direction finders such as those used by the Canadian Coast Guard for a number of years. (Radio direction finders allow a station to find the location of someone using a radio.) The admiral's response: "The NDRSMP already includes an integrated direction finding capability. Moving ahead with the NDRSMP is the best way to get an integrated direction finding capability to the field. To alleviate the current shortfall of the capability, this year the Coast Guard funded procurement of interim direction finding capability for those units with the greatest and most immediate need."[25] Admiral Riutta appears to have misunderstood the question: he did not respond to why the Coast Guard did not have a radio direction finder, as one had been in use by the Canadian Coast Guard.

In the June 1999 article in *Coast Guard,* Admiral Loy also wrote, "More importantly, a dull knife is a dangerous tool. Quillayute River taught us that we need to improve surfman training. It cost us three lives to learn the truth, but two years later, less than half our surfmen billets are filled by certified surfmen, and the average boatcrew experience throughout the Coast Guard has dropped to less than one year."[26]

When queried on his comments that the lack of surfman training resulted in the deaths of the three crew members at Quillayute River, Admiral Riutta replied, "The Commandant stated that we needed to improve surfman training. . . . The real issue is lack of surfmen, not the quality of training that surfman candidates receive."[27] There is no mention of poor surfman training in the published investigative report. In fact, Commander James H. Hasselbalch, the head of the investigation board, contended, "If a surfman had been on MLB 44363 that morning, the casualty would not have occurred."

One surfman remarked, "The Coast Guard will not accept any blame for mishaps. Because mishaps will continue to happen, as far as I am concerned, the Coast Guard is a dysfunctional corporation." Admiral Riutta retorted, of course, that "The surfman you quote is expressing a personal opinion about why he does not desire to go back to a surf station. While respecting his choice, we do not agree with his characterization of the Coast Guard as an organization that does not understand or support surfmen, stations, or the search and rescue mission."[28]

Lastly, officers have revealed that no one was blamed for the deaths in this incident. They carefully couch this revelation with the comment that "no blame" was attached in the mishap investigation. Those who know nothing about such investigations would therefore believe that there was no blame attached to anyone. Yet, the officers fail to mention the administrative investigation, which is the version that reached the media; this version put the blame squarely on Bosley. Admiral Robert Kramek commented, "As we can best determine, the motor lifeboat was not safely navigated, resulting in the loss of life for three of the four crewmembers assigned." Rear Admiral J. David Spade admitted, "I find the proximate cause of this accident was BM2 Bosley's failure to conform to the standard of care of a reasonably prudent Coast Guard coxswain."

. . . 18 Causes

What exactly happened in the first few hours of 12 February 1997 at Quillayute River? The following is a distillation of what many surfmen contend. Once BM2 David A. Bosley learned the *Gale Runner* was trying to enter Quillayute River on this stormy night, he did what any officer of the day should: he contacted the senior duty officer, BM1 Jon Placido. At this time Bosley apparently did not know the actual position of the sailboat. He knew that the radio transmission claimed the sailboat was at the bar and intended to come into the harbor. Placido instructed Bosley to obtain the location of the *Gale Runner.* Placido thought the problem might be solved by telling the skipper of the craft to turn into the seas. For some reason, Bosley had not informed Placido about the state of the weather and the National Weather Service's forecast.

While trying to obtain the position of the *Gale Runner,* the next audible transmission received at Quillayute River was the Mayday broadcast by Marcia Infante. Bosley had FA John DeMello sound the SAR alarm and then told Placido that the boat was taking on water; he was going out the door. Placido said he was on the way in and told Bosley to contact Master Chief LaForge.

Some have asked why Placido let Bosley get under way when he did not know the state of the weather. Many surfmen have explained that because the speed of a 44-foot motor lifeboat is slower than a helicopter, it is better to get a boat started right away. Starting the boat does not necessarily mean that the boat would sacrifice risk management. Those who have not served at a small-boat station do not understand how the adrenaline level rises when the SAR alarm rings. Very few can overcome that first rush to go immediately. The adrenaline level can be controlled once a boat crew is assembled and has started out. Then the rest of the information can be relayed, the crew completely briefed, and safety equipment, if needed, checked before going into any rough seas.

Bosley decided to get under way before Placido and LaForge arrived, and assumed control of what happened next. Master Chief LaForge testified that he radioed Bosley to go to the bar and see how it looked.[1]

Did Bosley proceed too quickly and did adrenaline seize him? Yes, on both counts. His running back to the station and yelling for his crew indicates a loss of control produced by an adrenaline rush. Master Chief LaForge's statement that he could not see the lifeboat's navigational lights indicates another sign of a hurried departure. According to Wingo's testimony, the three crewmen were in their surfbelts, but it is uncertain about Bosley. Why Bosley did not have the crew in their helmets is unclear. Bosley may have been too narrowly focused on the plight of the *Gale Runner.* The next series of events requires a considerable amount of speculation, and there is no way to know with any degree of certainty whether any of it is correct.

Bosley decided to cross the bar. Perhaps his reasoning for crossing was Marcia Infante's transmission that the *Gale Runner* was on the bar. Then the station received a Mayday broadcast. His public address system announcement of sailboat on the bar taking on water strongly suggests this reasoning. Bosley probably felt he would find the sailboat wallowing on or near the bar. This interpretation may have driven him to thinking he could help someone in distress; this motive would be called "moralizing" by Coast Guard headquarters. For reasons known only to him, Bosley at that moment felt he could handle the seas that faced him. No one in the small-boat community thinks Bosley would deliberately do something that would cause the death of his crew and himself.

Once committed to his course of action, Bosley's training took over. He steered the required heading of 210 degrees and started to look for the *Gale Runner.* If Bosley felt the sailboat was close to the bar, he might have thought the seas and wind would push the sailing craft close to James Island, hence his trackline close to the island. CWO2 F. Scott Clendenin has remarked that, "We train them to chase the white water," or, in other words, to keep your bow into the high waves. Bosley very likely was chasing white water, but was lured into thinking these were the true sea. Instead these were reflective waves. The administrative investigation stated that Bosley lost situational awareness and this may be true, but perhaps he decided to operate close to James Island, expecting to see the *Gale Runner* pushed close to the island. Bosley might have been fooled by the reflective waves, but when he neared the seaward tip of James Island the true sea struck the motor lifeboat broadside and started the fatal chain of events.

BM1 Michael Saindon, the current XPO of the Quillayute River station, offers another scenario. If one transcribes the radio traffic of those first few minutes, when Infante called that they were on the bar and planned to enter the river, and then looks carefully at the partial position she sent out, it is clear the *Gale Runner* was some three miles to the south of the bar. So, why did she indicate that they were at the bar, and why did Bosley stay close to the island? A navigational chart reveals the answer. People familiar with the area know to go to the entrance buoy—also called the "Q" buoy—and then start to make their approach to the river. Bosley and the other U.S. Coast Guard station personnel were trained in this method; it is based upon local knowledge. A person unfamiliar with the area, however, will look at the chart, see the range lines—marked "A" on map 4—and believe that the lines indicate the approach to the bar. The lines begin near The Needles, where the *Gale Runner* ran into problems. Infante and Schlag, glancing at the chart and having no local knowledge, would properly say they were at the entrance. Yet Bosley, and the rest of the station would presume the sailboat to be near the entrance buoy. Bosley, thinking a boat on the bar in the current weather conditions would be pushed toward James Island, may have elected to go much closer to the island than normal; thus his trackline. Mike Saindon related that in the summer of 1999 they had almost the same

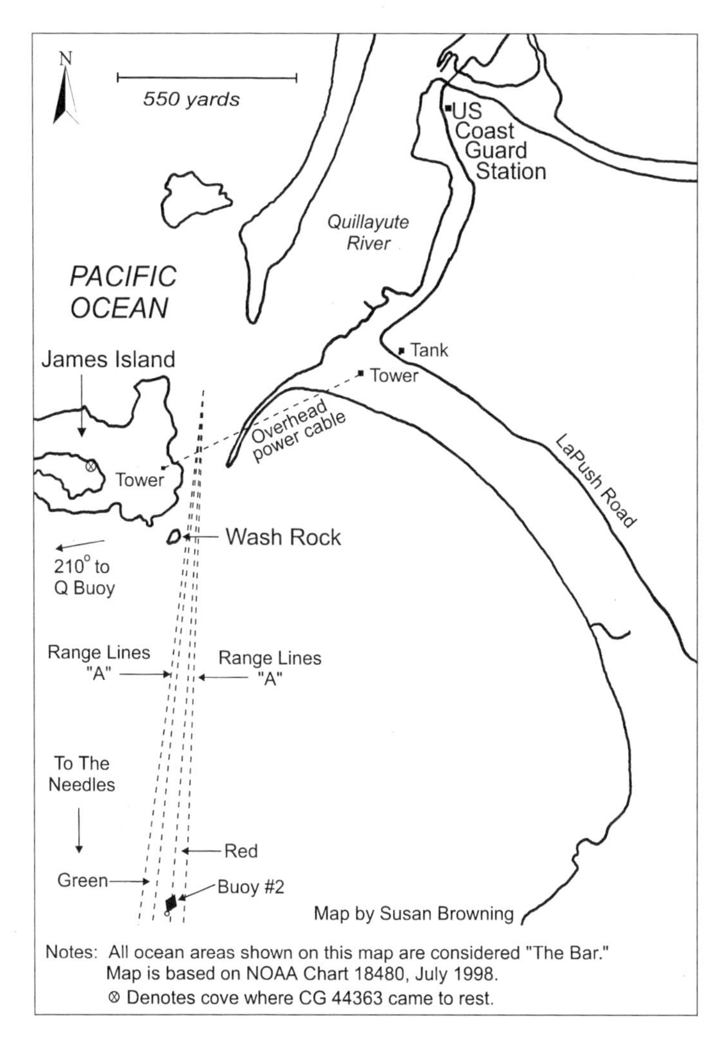

Map 4. The entrance to the Quillayute River.

type of call as Bosley faced that night in 1997, with two major exceptions. According to Saindon, "We received a call from a sailboat that said it was at the entrance to the bar and drifting toward the rocks. The big difference was it was calm and during the day. We got under way in the RHIB and went out to the bar in our normal approach. When we arrived on scene, no one was there. We looked to the south and there the boat was, near The Needles. The sailboat owner had looked at his chart and made the call using the chart as a guide." To prevent another similar episode such as what may have happened to Bosley, Saindon revealed that "We now train our crews that if you do not see what you're looking for when you first hit the bar, go to the entrance buoy—the Q buoy—and then start your search."

CWO2 Tom Doucette offers another scenario. "A while ago we went to open-eared helmets, away from the motorcycle type that covered our ears, but they expose the ear to the cold of the sea, which affects your equilibrium. Simply put, if you inject cold water into only one ear it introduces a spinning sensation. Cold water makes you feel like you are rotating away from that ear, warm water makes you feel like you are rotating toward that ear. In other words, if Bos was headed out to the southwest and got blasted with cold water in his right ear, he would have felt the boat was turning left. With no compass, no visual reference, no rudder angle indicator, he would naturally apply right rudder so the boat felt like it was going straight and it would turn toward the island. At Cape Disappointment and Grays Harbor, the big bars, you can get off course and be okay for a while. Quillayute River is not very forgiving of a few seconds of 'oops.'"

The first scenario above does not differ greatly from the published administrative investigation except that it shows Bosley may have had a plan, which the official report seems to refute. It also helps explain why Admirals Kramek and Spade both noted that Bosley, "for the best of reasons" decided to go to the rescue of the *Gale Runner.* This scenario also explains why Admiral Loy stated that proper risk assessment had not been practiced. Those who are removed from this case come to the same conclusion—thus the official verdict against BM2 David A. Bosley.

Yet FA John DeMello, in describing his ordeal on the beach, described the situation succinctly: "I learned that not everything is as it seems."

The U.S. Coast Guard's official verdict on the Quillayute River deaths, that "the motor lifeboat was not safely navigated," does not adequately explain what happened. The personnel policies of the U.S. Coast Guard and the way in which the small-boat rescue stations have traditionally been viewed and treated within the service have as much, if not more, to do with what happened than Bosley's presumably faulty decisions.

The only major changes immediately noticeable at the station level since the deaths have been changing the hooks on the surfbelts and exempting qualified surfmen from the sea duty requirement to make chief petty officer. Yet neither of these changes is valid. The new hooks on the safety belts have now been withdrawn. Several surfmen have revealed that the requirement for sea duty had been eliminated in the past, but then reinstated again. Officers-in-charge, commanding officers, and surfmen had all tried to have the sea duty requirement rescinded. In short, it took deaths to renew this logical policy.

The sea duty waiver was headquarters' only nod to personnel problems. In the December 1999 issue of the U.S. Naval Institute *Proceedings,* Captain Dana Goward, who is in charge of the U.S. Coast Guard's small boats, noted that "boat crew experience . . . has declined to less than 12 months on average."[2] The basic reason for the decline in boat crew experience resulted from the large drop in enlisted people beginning in 1995, as illustrated in table 1.

Admiral Riutta admitted that the cause of personnel problems is "our workforce shortage."[3] One apparent reason for the workforce shortage began under Admiral Robert Kramek. On 16 October 1995, Admiral Kramek, the Commandant of the U.S. Coast Guard at that time, issued an all–U.S. Coast Guard message calling for "streamlining." Admiral Kramek claimed that this order had occurred because of "a mandate to reduce the size of government without reducing service to the public." Admiral Kramek also admitted that he recognized "the changes are significant. In fact they represent about 25 percent of overall plan to reduce

Table 1. Officer and enlisted personnel, 1945–1999

Year	Officers[a]	Enlisted	Total	Ratio officers to enlisted
1945	12,902	158,290	171,192	1:12
1955	4,053	24,554	28,607	1:6
1965	4,825	26,832	31,700	1:6
1975	6,807	29,981	36,788	1:4
1985	7,508	31,087	38,595	1:4
1995	8,330	28,401	36,731	1:3
1996	7,908	27,129	35,037	1:3
1997	7,079	26,770	34,717	1:3.8
1998	7,945	26,945	34,890	1:3
1999	8,015	27,251	35,267	1:3

Source: Information Please Almanac (New York: Dan Goldenpaul Associates, 1969), 796, and *World Almanac, 2000* (Mahwah, N.J.: World Almanac Books, 1999), 212.

a. Numbers include cadets at the U.S. Coast Guard Academy.

4,000 personnel. . . . To date we have reduced nearly 3,200 personnel."[4] The rumor has circulated for years within the service that perhaps Admiral Kramek prosecuted too vigorously the mandate to downsize, while other government organizations supposedly dragged their collective feet without being punished. A New Jersey newspaper noted, "Some marine-safety experts and Coast Guard supporters contend the service should have resisted budget constraints that were forced upon them by deficit-reduction measures and other spending limits."[5]

By 2000, according to *Government Executive Magazine,* most of the current leadership in the U.S. Coast Guard admitted that the service reacted "overzealously and naively." "The critics may be right, but our cultural ethic forced us to play it straight," according to the current Commandant of the U.S. Coast Guard, Admiral James M. Loy. "We could have resisted streamlining and seized any number of opportunities to play politics or cash in on news events to increase our base during a strong economy. But we didn't."[6]

Another important reason for the workforce deficiencies is the policy of transferring nonrated personnel rapidly to service schools. Individuals out of boot camp can be transferred to a small-boat station, yet if they are not striking (learning their rate on the job) for boatswain's mate or machinery technician, they can be transferred to a school within nine months of arriving at the station. People unquestionably want to attend a service school to advance, and the service's shorthandedness makes attendance almost a necessity, yet the service does not provide enough time to have well-trained boat crews. After the deaths, Master Chief LaForge commented on the high turnover in the crew at Quillayute River. Official records support LaForge's remarks: of the 17 people at the Quillayute River unit in February 1997 who were normally assigned to boats (this does not include the officer-in-charge and executive petty officer), 12 had been at the unit for one year or less. Before the deaths in February 1997, only *five* people at the Quillayute River station had more than one year at the unit; 70.5 percent of the crew had one year or less of experience at a station in the dangerous waters of the Pacific Northwest. The station also had three surfmen rather than four.[7] Interestingly, the investigation into the losses in 1961 of five U.S. Coast Guardsmen who died going to the rescue of the fishing vessel *Mermaid* on the Columbia River brought out editorial comments about how the crewmen had little experience in the area. Apparently the problem of high turnover in the U.S. Coast Guard has persisted for 38 years.[8]

Two years after the deaths at Quillayute River, the personnel situation had improved somewhat. In February 1999, there were 22 people normally assigned to boat crews (not including the commanding officer or executive petty officer). Of these, 13–15 percent of those attached had one year of experience or less at Quillayute River. The station continued to operate with only three surfmen instead of four; the officer-in-charge and the executive petty officer were two of the surfmen.

In 1997 Vice Admiral James M. Loy, then chief of staff of the U.S. Coast Guard, pointed out the failure of the station to choose a proper experience level for the ready boat during the *Gale Runner* case. An examination of what was available that night reveals a clearer picture of the problem. Without pressures, it would indeed be conceivable that Master Chief LaForge and BM1 Placido would drive into the station, then look at the roster, select seven crewmen for the two boats, recall the needed personnel, and then put out to sea. Because there were only three surfmen attached to the unit, if both LaForge and Placido had gotten under way as surfmen, the command and control structure at the station would have been gone. Therefore, one would have had to remain at the unit until the other surfman returned to the station.

In the section on duty that night, Bosley could have gone on the first boat as a crewman even though he was not a surfman. Wingo and Miniken would not have gone; they had less than a year at the unit. Schlimme would have gone. No one else in the section on duty had over one year's experience at the unit. So, at least one crew member on the ready boat would have had to be recalled from the other duty section. People from the other duty section would also have had to be recalled for the standby boat. In other words, there were not enough people at the station with over one year of experience to make up two full boat crews per duty section. In this scenario, remember that while the boats were waiting for off-duty crew members to show up, someone had put out a Mayday call telling the station that they are taking on water. To maintain the level of readiness that the chief of staff seems to have expected of the unit, everyone stationed at Quillayute River would have had to remain at the station at all times.

When questioned about the high turnover rate at Quillayute River, Admiral Riutta wrote, "the tour length at this semi-isolated station is 3 years due to the isolation. Therefore, at least one third of the personnel will leave each year by design. Non-rated personnel departing to attend 'A' schools cause more turnover, and we are currently examining ways to reduce this category." The admiral also reiterated what he had previously stated: "we are still able to meet minimum mission requirements by modifying watch rotations, training schedules, etc., in order to maintain at least minimum mission readiness at all stations. If a station OIC, commanding officer, or anyone else in the chain of command determines that a station is marginal or below minimum conditions of readiness, we work together to either solve the problem or have the unit stand down until readiness is reestablished to an acceptable level."[9]

In 1991, a U.S. Coast Guard station staffing study indicated that "The central problem identified in this study is that many of our personnel arriving at the stations are unqualified to fill the billets to which they have been assigned. Existing station standards are based on the assumption that assigned billets are filled with qualified personnel."[10] The study also mentioned the number of hours people at stations worked. Yet, nine years after the study, both problems remain.

Admiral Riutta's response admitted that the

> current staffing standard is based primarily on operational workload, predominately SAR time on sortie. This standard does not adequately capture the array of work being done at stations. The Center for Naval Analysis is currently developing staffing standards for Groups, Aids to Navigation Teams (ANTs) and stations that will encompass the full range of station workload, including training. The standard will provide for sufficient billets to offset the training load currently on small boat stations. Similarly, once stations are staffed with sufficient personnel to address the full range of workload, workweek hours will decline. The Coast Guard Personnel Command makes assignments in part based upon qualifications. Many members develop skills necessary for qualification during previous tours, however, each unit has unit specific qualification requirements related to the area of responsibility. Our "Personnel/Training" system is not able to provide off-unit/advance training for every assignment. It is not a perfect system. Couple these facts with the Coast Guard's personnel shortages, increased retirements, lack of accessions into the enlisted workforce, increased officer accessions from enlisted ranks and you have a very challenging workforce problem.[11]

Admiral Riutta still does not mention how the stations will be staffed if they are unable to enlist enough people.

Table 1 vividly illustrates that the rapid decline in the enlisted force is key to what happened at Quillayute River. One of the primary reasons Bosley remained at the station is that both Senior Chief Shipman and Master Chief LaForge knew that the shortage of personnel would make it very difficult to find a replacement. After Bosley's death, BM2 A. J. Scholtz, his replacement, reported aboard in August, five months later. Scholtz soon was promoted to BM1 and was transferred by May 1998; his tour at the unit lasted only nine months.

Another personnel policy that causes problems at the stations is the service's tendency to transfer people in the summer. This policy prompts instability at a station at traditionally the busiest time for search and rescue at a unit. Apparently headquarters anticipates that people coming from other units are immediately qualified to begin handling SAR cases. Yet station commanders point out that rarely do people coming from ships have the qualifications to begin immediately handling cases. Even if the person comes from another small-boat station and is fully qualified, the person must learn the new area. One reason for this policy is that planners at headquarters have no idea of what takes place at a small-boat station. Admiral Riutta insisted that "[e]nlisted personnel are not all transferred during the summer. Our personnel policies are designed to balance the needs of the service and the needs of our personnel. Our normal policy is to minimize transfers during the school year, but when the service needs dictates otherwise, we proceed accordingly. It is up to the unit OIC or commanding officer to make that determination and we make that very clear to them. The timing of transfers is a major quality-of-life issue for our personnel and it is a key retention factor. Two years ago, however, we expanded the normal rotation window from a May to September timeframe to one which extends from March through October in order to reduce the problems caused by turnover, particularly at our smaller units such as stations and patrol boats. We also monitor unit readiness and take whatever action is needed to at least meet the minimum service needed."[12]

The comment about "expanding the normal rotation window . . . to one which extends from March through October" implies that the normal summer rotation schedule does impact the stations.

Another issue that contributes strongly to what happened at Quillayute River is the way in which the service views its small-boat stations, as illustrated by an old saying, "small boats, small problems." Many in leadership positions feel that duty at the boat stations is easy and that the units should be used as a reward and resting place between arduous duty on cutters. The rescue units most identified with maritime rescue in the United States appear to be held in little esteem by the leadership of the U.S. Coast Guard, an observation that requires some explanation.

The desire to help those in distress upon the sea is an important part of our maritime history. Saving lives from shore-based stations began in earnest as early as 1802, when the first lifeboat station was erected in Cohassat, Massachusetts. During the earliest years volunteers served at small, shed-like stations. This setup did not succeed very well on the national scene, and in 1878 the federal government established the U.S. Life-Saving Service.

By 1872, as Sumner I. Kimball, who would eventually become the general superintendent of the U.S. Life-Saving Service, laid the groundwork for the organization, he detailed Captain John Faunce of the U.S. Revenue Marine (as the U.S. Revenue Cutter Service was then called) to inspect the current lifesaving stations. Faunce's report helped Kimball shape the service. The U.S. Revenue Cutter Service—formed in 1790—had a commissioned officer corps. Kimball eventually had the U.S. Revenue Cutter Service assign officers, usually lieutenants, to the U.S. Life-Saving Service as inspectors. A captain assumed control of the inspectors. In general there was a good working relationship between the two services, but the U.S. Life-Saving Service, a civilian organization, remained in charge.

By the beginning of the 20th century, however, the U.S. Life-Saving Service faced many difficulties. It used outdated equipment and was gradually moving into the use of motor lifeboats. It had difficulties recruiting personnel. It had no retirement system and no compensation in the event of injuries received in the line of duty. According to one student of the service, it was not uncommon to find 70-year-old men on the sweep oar and 60-year-old men at the stroke oar.[13]

Then there occurred another of the many efforts to "streamline" the federal government. In a move to consolidate a number of small federal maritime entities, the U.S. Life-Saving Service and the U.S. Revenue Cutter Service merged in 1915 to form the U.S. Coast Guard. The reorganization brought military ranks for all and a retirement system. Over the next 40 years, for all practical purposes the U.S. Coast Guard remained divided into two groups: those who served aboard cutters and were commanded by commissioned officers (the old U.S. Revenue Cutter Service), and those who served at shore-based small-boat rescue stations commanded by enlisted men and a smattering of warrant officers (the old U.S. Life-Saving Service). Even though at the time of the merger the U.S. Life-Saving Service had more personnel, the other service became the dominant group, because they could boast a commissioned officer corps who had trained at an academy and were considered educated officers and gentlemen. The other group was perceived to consist of uneducated watermen. Those in the service's leadership, however, realized the value of the small-boat stations when it came time for budget appropriations. There is nothing like an amazing rescue in which the sea is narrowly cheated of human lives to help shake money loose from Congress for the entire service, which did not trickle down to the stations. Other than a few inspections, the cutter force left the stations to their own devices, giving them very little money. Those at the stations accepted that they would receive very little in the way of appropriations and publicity from the cutter branch of the service. The station personnel learned how to cobble together their equipment and keep it running despite low funds, and they took a fierce pride in being able to respond at all times to those in peril upon the sea.

During Prohibition, the U.S. Coast Guard expanded to wage the rum war at sea, only to receive major cuts after Prohibition's repeal. In 1936, Russell R. Waesche Sr. became the Commandant of the U.S. Coast Guard and shortly thereafter sought to eliminate a number of the small-boat stations. Admiral Willard J. Smith, commandant from 1966 to 1970 and previously aide to Admiral Waesche, noted in his oral history that at the time the leadership felt it had to convince people they were sailors. Many officers believed that the people on the small-boat stations would make them a "laughingstock."[14] The small-boat-station people were not considered by the officer corps to be sailors. In this view, the leadership perpetrated the myth of the small-boat stations: the lifesavers served forever at a particular unit. A 1994 book on the U.S. Life-Saving Service, however, points out that on Lake Michigan, in the heart of many small-boat stations, the average length of service for the man in charge was eight years.[15]

It is important to understand that the small-boat stations of the U.S. Coast Guard have always been an enlisted man's program. For many years, only master, senior, and chief petty officers were in charge, with a smattering of warrant boatswains, who came from the enlisted ranks. The stations drifted along, in what one senior officer has dubbed "benign neglect," until the 1970s. The end of the Vietnam War and an increase in commercial and sport fishing, combined with an explosion in recreational boating, brought about a crisis at the small-boat stations. Crews at the stations were working extremely long hours. Officers were concerned about the retention rate within the service, as overworked people usually refuse to reenlist.

Officers began instructing chief petty officers that they could not work their crews so much. The chiefs replied that they had to work the crews the hours they did and that it was a safety issue: if they did not work these hours, people would die. Finally, headquarters acquiesced, agreeing that if shorter hours were needed, more billets must be available for the units. The number of people at a unit increased. Some stations reached 40, 50, or more people. Then the officer corps decided that the stations had too many people for enlisted staff to command; a commissioned commanding officer needed to be in charge. Some lieutenants began to take the helm, and a great many warrant officers were placed in command of the units.

Meanwhile, the stations were given more missions, beyond search and rescue, such as environmental protection. The U.S. Coast Guard's study of the stations in 1991 revealed that crews at the units worked over 100 hours a week. The study called for even more people for the units. During the service's last major cut in enlisted personnel, in 1995, the recommendations and study were shelved.

It has been inevitable since 1915 that there would be a crisis involving the small-boat stations because of two factors that have always been present within the U.S. Coast Guard. First, the academy officer corps controls the service. A basic tenet of the officer corps is that the organization is a seagoing service: everyone must go to sea, and small-boat crews are not sailors. Second, within the service, if a career path does not have options for academy officers, then it is looked upon as little more than useless. There is no career path for academy officers at small-boat stations.

In a budget crunch, therefore, the small-boat stations are the first to be cut. The 1990s brought the U.S. Coast Guard to such a crunch, and the reaction of the service to this budget problem relates directly to the Quillayute River incident. In the 1990s, when ordered by the executive branch to reduce their force, Commandant Admiral Robert K. Kramek gutted the enlisted force. The enlisted force dropped from 31,087 in 1985 to 28,401 in 1995, and then to 27,129 in 1996. In 1995, however, the officer corps did not decrease; it actually increased, not decreasing slightly until 1996.

Another factor involves the way in which the service managed their enlisted force prior to the Quillayute River incident. Previously, a petty officer who wished to advance had to have sea service. These rotation policies made it extremely difficult to obtain surfman qualifications. Qualifying for surfman could take the average person three years of work at a unit. Once qualified, such people should logically have been sent to another small-boat station to pass along their knowledge. Yet what actually happened was that they had to go to sea in order to be promoted. If they were not promoted, they would be forced out by an "up or out" personnel policy.[16]

Another reason given for reducing the stations was the belief that technology will cure all problems. Officers contend that with the new 47-foot motor lifeboat, plus helicopters, fewer stations and fewer small-boat hulls are needed. Headquarters briefers point out that with a speed of 25+ knots, the new 47-foot motor lifeboat can arrive quickly on a scene; thus some stations can be put out of service. Furthermore, the quick arrival time means that the service can reduce the number of boats, and thus needs smaller crews to staff stations. Boats, however, generally do not travel at 25+ knots into a heavy sea, the condition in which they are most needed, nor can they usually tow a boat at the motor lifeboat's maximum speed. Another contention is that helicopters can go faster than motor lifeboats. Helicopter crews of the service have accomplished some amazing rescues, as demonstrated by the rescue of the two people from the *Gale Runner*, but the machines also have limitations. More than once something has happened to stop a rescue hoist, such as a broken hoist cable.

Limited knowledge about the small-boat stations has prompted seemingly irrational decisions. Stations have been closed, only to be reopened a short time later (most reopenings can be attributed to political pressures). Moreover, recent plans for the use of the U.S. Coast Guard Auxiliary to help the stations by working in the surf illustrate the lack of knowledge about the inherent dangers of the work. Despite the contradictions, the problems continue and the reason, according to one young academy officer, is that "no one wants to tell the boss the bad news." Implied in his statement is the belief that bad Officer Evaluation Reports (OERs) can result from criticizing the organization.

In the small U.S. Coast Guard, a bad OER can end a career. As retired Captain Philip C. Volk has pointed out, "Once you make a mistake, that's it, no second chances. It is a service of no second chances. Excellent isn't good enough in the Coast Guard anymore." To insure that an officer does not have anything negative in his or her OER, there is a tendency to pass regulations to make sure nothing does go wrong. As Captain Volk commented, "What happens then is [that] it is safer to do nothing."

More importantly, however, the problems and duties of small-boat stations are outside the experience of the service's senior leadership. The enlisted people who are dedicated to the small-boat rescue stations have a sense of history, tradition, and esprit de corps concerning the stations that the service's officer corps has not experienced. All the risks and all the dangers inherent in small-boat operations are taken by the enlisted force, who become the experts. All the policies affecting their lives, and the lives of civilians upon the water, are set by an officer corps that has little experience in the field. While one may argue that senior officers at the headquarters of the U.S. Marine Corps or U.S. Army are as far removed from their troops in the field as officers in the headquarters of the U.S. Coast Guard, marine or army officers began at the platoon level, in the mud with their troops, and faced the same dangers. Apparently no one at U.S. Coast Guard headquarters in senior policy-setting positions has ever served at a small-boat station.

The system could work, *if* those in senior positions would consult with and listen to those in the field—the experts. This consultation had not been happening prior to 12 February 1997. The communications gap—chasm is a better word—between those who serve at the units and those at headquarters is amazing. One senior officer commented, "It is ironic the stations think headquarters is not listening to them." A junior officer explained that "headquarters is listening to them because they granted their request to be in charge of the units." This type of thinking emphasizes the communications gap between headquarters and the stations.

The feelings of those at the stations are akin to the bitterness of the troops in the trenches toward staff officers during World War I. Those doughboys, poilus, and Tommies of many years ago felt staff officers sent them off to die without caring for them, knowing anything about them, or visiting them.[17]

One way to assess how a bureaucracy feels about the various divisions within the organization is to examine how each division is funded. The small-boat stations do the majority of the search and rescue in the U.S. Coast Guard, but search and rescue ranks fifth in the service's budget. Surprisingly, the stations must share that budget with funds for all search and rescue within the service. In the U.S. Coast Guard's budget for fiscal year 2000, the following ranked higher than search and rescue: drug interdiction (17.72 percent), aids to navigation (15.50 percent), living marine resources law enforcement (15.39 percent), marine environment protection (11.44 percent), and finally search and rescue (11.50 percent).[18]

Admiral Riutta commented on this matter, remarking that "[a]lthough the SAR program ranks fifth in terms of the percentage of the Coast Guard's overall budget, analyzing the Coast Guard's budget does not provide a sufficient basis to draw a conclusion about the prominence of the SAR mission in the Coast Guard hierarchy. Search and Rescue is a primary mission of all Coast Guard units. As a multi-mission organization, we respond to the highest priority mission—the safety of life at sea. We do not subscribe to your conclusion that search and rescue is not an organizational priority, because of the funding the program receives relative to other programs."[19]

In spite of Admiral Riutta's reply, not all U.S. Coast Guard units appear to have search and rescue as a primary mission. For example, the U.S. Coast Guard headquarters official web page on the new cutter *Healy* declares that the cutter's "primary mission includes functioning as a world class high latitude research platform." The cutter is "employed in icebreaking operations during all seasons in the Arctic and Antarctic. All ship systems are designed to function for extended winter operations in these areas including intentional wintering over."[20]

Other units, such as merchant marine inspection offices, rarely, if ever, have search and rescue as a primary mission. A history of the Tampa Marine Safety Office, for example, states, "the Marine Safety Office in Tampa . . . [watches] over the seventh largest port in the United

States and keep[s] the traffic flowing smoothly as Captain of the Port. Other duties included licensing merchant marine seamen, inspecting marine vessels, investigating marine casualties and monitoring the cleanup of oil or chemical spills, all part of the current Marine Safety Office."[21] Although search and rescue is not explicitly defined as a major portion of its mission, the program that merchant marine inspection offices operate under receives more funding than search and rescue.

Another contributing problem is the U.S. Coast Guard's inability to define itself and its willingness to be all things to all people. According to one retired officer, the organization is always "chasing butterflies." The U.S. Coast Guard seems to search endlessly either to "find" itself or to "save" itself. The pages of the U.S. Naval Institute *Proceedings,* the professional magazine for the federal sea services, illustrates this searching for identity or a course of action that will save the service. In the October 1997 issue, for example, Lieutenant Christopher Forando argued that unless the service built a special-operations force and aligned itself more closely with the U.S. Special Operations Command, it "faces becoming a search and rescue and marine safety-oriented service."[22] The tone of the article indicates that search-and-rescue work is less important.

Every 10 to 20 years the U.S. Coast Guard tries to redefine itself. In the 1960s, oceanography would save the service, and the organization would take the lead role in the federal government in studying the sea. Every major cutter, including the overworked buoy tenders, did some type of oceanography. A new enlisted rate was formed. An oceanographic ship even made it to the design stages. Then, without explanation, oceanography became passé. In the 1970s, the scene shifted from below the seas to the surface, and the U.S. Coast Guard focused on becoming premier seagoing police officers. In the 1980s, the service directed its efforts to the environment. The 1990s brought "Deepwater," which plans to replace the service's aging cutter fleet and aircraft. Everything else apparently must be put on hold to save the cutter force. Throughout this attempt to define its mission, the service has not emphasized search and rescue. Meanwhile, the people who served at the small-boat stations worked long hours, went shorthanded, and lacked equipment and repairs. It is now past time for someone to say "enough." Of course, such a pronouncement is not career-enhancing, but it is time to help crews, not careers.

It can be argued that oceanography, law enforcement, and environmental protection missions are missions deemed important by either the executive or legislative branches of the government; these additional missions have also brought officers the chance for rapid promotions and new sources of funding for the service.

At the beginning of the 21st century, the service finds itself in a tough position. It is operating with too few people, a factor caused by misguided planning and strict budget constraints. Moreover, the U.S. Coast Guard has too many missions. Even so, the executive and legislative branches of the government, largely for political reasons, continue to task the U.S. Coast Guard with a myriad of duties, including search and rescue from small-boat stations.

After the deaths at Quillayute River, the Coast Guard began using a new lexicon. According to Admiral Riutta, search and rescue is the service's primary mission. But according to the new more complicated lexicon, "By law the CG [Coast Guard] is mandated to be ready to conduct search and rescue missions but we are also by law not required to assist, we 'may render assistance.' As such, the CG operates under an us, ours, them, theirs priority for setting out to sea for rescues."[23]

Another example of the new thinking is that some stations in the Thirteenth Coast Guard District must now broadcast on marine frequencies that "environmental conditions currently exceed the operational limitations of [a station's] resources. Response may be delayed as a result. Mariners are advised to exercise extreme caution when operating in the vicinity [of the station's area of responsibility.]" Headquarters has also "discovered" that the people at the stations are working very long hours. This discovery required yet another study. Apparently, no one knew about the 1991 staffing study that stated people at the stations worked long hours. The new study, not too surprisingly, found that if a person works too many hours he or she can make mistakes. The commandant shortly thereafter issued new "fatigue standards" for all the stations. Henceforth, officers-in-charge and commanding officers must not work their crews more than 68 hours a week.

One Thirteen District briefer said that the station at Quillayute River, after working so many hours, would "close the doors." Because of the fatigue standards, the station watch schedule now follows this routine: at 6:00 P.M., the only people who remain on the station are a communications watchstander and one backup person. The rest of duty section takes a government vehicle and goes home, awaiting any call. Some have at least a 20-minute drive. The only time the duty section does not go home is if the weather is bad—not the weather on the bar, but the weather on the road. This plan is approved by the district.

Admiral Riutta has written, "New fatigue standards are being developed in an effort to ensure the safety of our boat crews. Our people are our most important resource. A fatigue standard, coupled with an accurate staffing standard, will enable stations to work better with less risk due to sleep deprivation. The fatigue standard under development does not state that after 68 hours stations will stop responding. It does stress that it is important to recognize and respect the limits which we can work our crews."[24] The service, at present, does not have an accurate staffing standard. What is taking place is a piecemeal plan that does not address the problem: a lack of people.

While Admiral Loy has been noted within the U.S. Coast Guard for his appreciation of the history of his organization, these actions contradict the mission of the service. At the time of writing of this manuscript, the new standards have not yet filtered down to the public.

New policies and rules can have an adverse effect. For example, if the person in charge of a station has not followed the fatigue standards as conceived by headquarters, the district, and the Group, then headquarters can claim that the person of the unit did not follow the standards; thus any death that might occur can be blamed on that person. If the person has followed the standards, then headquarters can exclaim that it has never been the Coast Guard's intent for a station to shut down; again the commander of the unit would be at fault. Joseph Heller wrote a best selling novel, *Catch-22*, based upon just such an arrangement in the military.

. . . 19 Lessons Learned

What did the Quillayute River deaths teach the U.S. Coast Guard's leadership? Everything to emerge from the investigations has apparently been aimed at modifying existing regulations or equipment. The waiver of the sea duty requirement for advancement from petty officer first class to chief petty officer is an example.

Admiral Loy, however, has stated that following "the tragedy, we developed 38 action items to improve safety and operations throughout the small boat community. The action items ranged from reviews of standard operating procedures to training and electronic tracking of our small boats. Only one of these items has not been implemented . . . and that item is under consideration as part of our National Distress and Response System Modernization Program (NDRSMP)."[1] Admirals Loy and Riutta, however, only listed one of the "38 action items." One safety officer admitted that he did not know there were "38 action items" developed "to improve safety and operations throughout the small boat community."

When the results of the official investigation were released, Admiral Kramek ordered the service to use the accident as a training scenario for team coordination training (TCT). The scenario is now also used at the U.S. Coast Guard Chief Petty Officers Academy and leadership training school. A request for class material from the Chief Petty Officers Academy and the leadership school—both located at the U.S. Coast Guard Academy—resulted in a response from the former, but not the latter.

The two training scenarios received are interesting. The oldest, and closest to the date of the accident, is dated December 1997. Although the most recent is not dated, it appears more contemporary. Each of the case studies has sections that cover a synopsis, damage and injuries, factual information, causes, and lessons learned.[2] In the earlier scenario, the synopsis is fairly consistent with the published information in the administrative investigation, although it mentions that Seaman Wingo was assigned to look for the lighted daymarker no. 3, recorded as buoy 3 in the investigation, and Wash Rock. Neither of these two comments is revealed in the testimony given in the investigation. Moreover, there is no reference to the station's policy that a surfman did not have to be aboard if the bar conditions were not threatening. What is mentioned, however, is that the executive petty officer and the officer-of-the-day (OOD) went to the bar during daylight and saw the bar in a relatively good condition. Next, the case study contends that at 2200 hours [10:00 P.M.] the OOD called Placido, saying the winds had risen a little but conditions still were not threatening. The study does record a decision not to have a surfman aboard because of "an inaccurate estimation of the weather conditions by the coxswain [OOD] who did not completely convey reported and forecast weather conditions."

One item in the factual section of the scenario is revealing. After MK3 Schlimme, the boat engineer, made a comment about getting out of here, according to the training scenario, BM2 Bosley then said, "'Screw this,' 'gunned it' and went into the seas." This statement had not been described in the administrative investigation.

Albeit a minor discrepancy, in both the 1997 and newer scenarios, the U.S. Coast Guard is apparently concerned about using the profane language from the official record; there is none in either scenario. A not-so-minor item is the omission of Master Chief LaForge's transmission to Bosley that he had not seen a sailboat near the bar and to proceed to the bar to look things over. Interestingly, in both the 1997 scenario and the newer case study the causes and lessons learned are just a repeat of the 1997 published report by Vice Admiral Loy, at the time the chief of staff of the U.S. Coast Guard.

In the more current case study, when the first motor lifeboat got under way, "The engineer placed a boat crew safety belt on the coxswain while he was still sitting in the coxswain chair as the MLB [motor lifeboat] was on the way out to the Wash Rock on the island. The coxswain seemed to be upset that the engineer put the boat crew safety belt on him." Nothing within the administrative investigation reveals this piece of information.

There are two possibilities as to where the case studies secured information not in the administrative investigation. Whoever wrote the study might have decided that this was probably what happened. More likely, however, the material came from the safety investigation or the Commandant's Vessel Safety Board, both of which are classified.

It is interesting that the U.S. Coast Guard's leadership is apparently basing studies on confidential information, yet also releases this information. This situation is doubly interesting for those trying to obtain information on the incident through official U.S. Coast Guard channels. An attempt to acquire Vice Admiral Loy's lessons learned, which is classified as "official use only," through a Freedom of Information Act (FOIA) request, resulted in a U.S. Coast Guard headquarters response that no copies of such a publication could be found. This response suggests that headquarters wants it known that no such publication exists, even though the training scenarios contained information from the "nonexistent" publication. As was the case with many research items for this book, as soon as informed people in the field learned of the difficulty of obtaining information from the publication, copies of it arrived from other sources. It seems that U.S. Coast Guard headquarters, with hundreds of people assigned to it, cannot find information they publish, whereas field units with many fewer people working extremely long hours can find such information. This situation vividly demonstrates the difficulties of obtaining sensitive information from headquarters.

The disconcerting issue about these training scenarios is their omissions of important items, such as the station's regulations and the wrongful impressions presented about Placido and LaForge. In a small service, this perception quickly spreads throughout the organization, fueled by its continual reiteration in training sessions. The worst aspect of the aftermath of the deaths of Bosley, Schlimme, and Miniken is how Placido and LaForge have been made to feel about something that is the result of the inattention of headquarters to the small-boat stations and to the gutting of the enlisted force.

Could something like Quillayute River happen again? On 25 August 1997, Adam Katz-Stone, writing in the *Navy Times,* discussed this question with CWO2 Randy Lewis, commanding officer of the Grays Harbor, Washington, station and BMC Tom Karczewski, his executive petty officer, along with CWO2 F. Scott Clendenin, commander of the Yaquina Bay, Oregon, station. The responses by these three senior people are revealing.

Katz-Stone wrote, "The people who run the 44-footers every day say the events that claimed the lives of three of four crewmen at Quillayute River . . . could unfold in exactly the same way tomorrow."[3] The senior people pointed out that the judgment calls made that February night are made every day. Could new regulations prevent this from happening again? The surfmen "doubt whether such a regulation could be written." CWO2 Clendenin remarked, "If anybody tells you we will never have another accident like that again, that would just be untrue." According to CWO2 Randy Lewis, "The hardest thing for us to get across to people is that our entire world is one of judgment. We are constantly in the process of evaluating information and making judgments.

"There is a fine line between regulating judgment, and removing the ability to perform. Every time you come up with a hard and fast rule—don't do the following—I guarantee you there is a situation where you need to do that thing."

CWO Lewis continued, "There were mistakes made. But none of it was so far out in left field. We operate on that exact same edge very frequently." On Bosley's not passing on the forecasted weather conditions: "I can't tell you how many times . . . [the National Weather Service] have projected that gale warnings would be up, and they didn't occur. Likewise, they can forecast a gale warning, and we may even have a gale blow through, but that might not even cause the bar to bump." BMC Tom Karczewski noted that normally the weather service provides "fairly good" forecasts, but "we all know things can come up unexpectedly, or ahead of time, or not at all."

Why did Bosley decide to push out into the seas? Lieutenant Martha Laguardia, at the time the Thirteenth Coast Guard District's public affairs officer, repeats the approved statement: "He was moralizing the urgency of the situation and therefore deciding to operate in conditions that were beyond his qualifications and above his capability." But CWO2 Clendenin claims that "It's hard to describe, when that search and rescue alarm goes off, that feeling that gets into your stomach. Sometimes you can forget to do things, or even forget what you have done. I'll talk to people in the debriefing afterward and they won't even remember what they did."

Can headquarters make rules that cover what is considered "rough weather?" There is some doubt in the minds of many surfmen. "The hardest thing to do is to put numbers on something that is a judgment call. What's a rough bar, anything over eight to ten feet? But sometimes eight to ten feet is beautiful, a long gentle swell," said Chief Karczewski, "where other times it can flat annihilate you."

Can something like Bosley's error in judgment be prevented in the future? Lieutenant Laguardia again gives the approved statement: This can be avoided by "sticking by the standard operating procedures, communicating with those people that are in your chain of command, and not trying to do more than you're capable of doing." Laguardia, however, does not mention that perhaps Bosley felt he was doing something he was capable of accomplishing.

Chief Karczewski admitted, "I'm really surprised that something like this hasn't happened before, given the amount of times that we have to go out in nasty lousy weather. We're taking steps to prevent it, but could somebody make the same mistakes again? Absolutely!"

CWO2 Clendenin aptly sums up what the Quillayute River deaths demonstrate: "What we do every day is very, very dangerous."

20 The Light at the End of the Tunnel?

The office in headquarters that most closely works with the small-boat rescue stations is Boat Forces. In charge of small boats up to 65 feet in length is a captain in the aviation field. At the time this book was written, Captain Dana A. Goward headed the office. Soon after taking over his duties, he started "Project Kimball" to "improve the integrated interrelated systems that constitute the boat force (groups, stations and aids to navigation teams [ANTs])." Furthermore, the project would "identify the issues and problems affecting mission performance, develop solutions, and effect beneficial change."[1]

Captain Goward insisted that "recent operational failures, such as the Quillayute River mishap and the loss of life in the 'Morning Dew,' 'Florida Air' and 'Miss Renee' cases, as well as the failure of support projects . . . have highlighted a general loss of focus by leadership and management on the readiness of our boat force." Echoing what many at the small-boat stations have said, Goward observed, "a plethora of studies have examined various aspects of the boat force, and made numerous recommendations. Despite these efforts, only marginal improvements have been made. . . . Consequently, these forces have been increasingly operated and supported on an ad hoc rather than a disciplined and business-like method."

For the first time in years of research on the small-boat rescue stations, it appeared that headquarters actually seemed interested in listening to what the people at the stations have been trying to tell the leadership. Could this be the light at the end of the tunnel?

Soon thereafter, an article appeared in the U.S. Naval Institute *Proceedings,* entitled "Life-Saving Service Left in the Cold," written by Captain Goward. The article describes the problems found at the small-boat stations, and in it he admits that the stations are "an enlisted operation competing for resources, management, and senior leadership attention in an officer-dominated organization."[2] He also states that the small-boat rescue stations "are at the heart of our corporate and public identity—they are the very soul of the service."[3] This is what enlisted people at the stations have been arguing for years.

Captain Goward's solution? Add officers to the groups and stations. As he puts it, "We need to provide [the stations and groups] with a constant, high level of leadership and management attention at every level in the organization. The only way to ensure that this is done on a consistent, long-term basis is to allocate a greater portion of our mid-level and senior leadership and management—our officers—to the task."[4]

What at first glance seems to be a new turn of reasoning on the part of headquarters appears upon closer examination to be an effort to create more billets for the officer corps. Under the subheading of "Recognition as a Professional Discipline," Goward writes, "Boat use, unique maintenance issues, crew fatigue, in-depth community involvement, reserve and auxiliary use, the service's most junior and inexperienced crews, and a plethora of other challenges define a unique and extensive group/station skill set. Success in this field depends on officer experience and expertise."[5] This statement insinuates that all the work done by the small-boat stations since 1878 has been accomplished by unprofessional crews. Yet, if so, how can the fine worldwide record the United States has for lifesaving from shore be explained? Furthermore, how has the Coast Guard done so well with chief petty officers and chief warrant officers as officers-in-charge and commanding officers of some cutters? William D. Wilkinson, director emeritus of The Mariner's Museum, Newport, Virginia, and an internationally recognized expert on the small craft of the

U.S. Coast Guard, has said that in his "many conversations over a long period of time with officials of foreign life saving services he has heard nothing but the highest praise and admiration for the work of the U.S. Coast Guard in its life saving activities from shore based stations."[6] Goward's comment contradicts that perception—it insists that officer presence validates the professionalism of the U.S. Coast Guard.

Captain Goward has rightfully mentioned the work the stations contribute to the service's search and rescue efforts, yet his article hedges the statistics somewhat. Goward claims that Groups (which have officers attached) and stations save 70 percent of the lives and 90 percent of the property. According to the published search and rescue statistics—the last available year was Fiscal Year 1993—the stations did 53.8 percent of the search and rescue, while groups did 19.1 percent and air stations accomplished 10.3 percent. Current figures should continue to demonstrate this disparity between the stations and other types of units; stations still carry the search and rescue load within the service.

Within the narrative of Captain Goward's text is a quote from a study undertaken by the U.S. Coast Guard in 1989: "Stations are given the organizational cold shoulder . . . they are appendages that the rest of the [Coast Guard] takes little time to understand. They are our grassroots, but they are not part of the mainstream . . . they get the job done in spite of our ignorance. This is the real problem."[7]

In his background material for the charter to Project Kimball, Captain Goward states, in part, "the loss of life in the 'Morning Dew,' 'Florida Air,' and 'Miss Renee' cases" have highlighted the "loss of focus by leadership and management on the readiness of our boat force." When asked to further explain what he meant, Captain Goward noted that the *Florida Air* "was a vessel that sank off the Florida panhandle out of the range of [Coast Guard] radio reception. It was mentioned because the new coastal radio system that will allow us to hear our customers and will be the backbone of the boat force's command and control was delayed many years (perhaps 5+) for lack of a sponsor/advocate in [headquarters] and the budget process."[8]

Captain Goward, in responding to a question that stated the inclusion of the *Miss Renee* case seemed to imply "that the stations did something wrong in . . . the case"; he wrote:

> No implication there, though you touch a point that I have been trying over the last 9 months or so to address, *i.e.*, there is a big difference between a flawed system that doesn't do what it should and flawed people who don't do what they should. In most cases where we (the Coast Guard) fail, it is not that our people have failed. In fact, our people typically perform heroically and admirably, especially considering the flawed system in which they are forced to operate. Our operational failures can usually be attributed to system flaws or failures that the excellence and heroism of our people were not able to overcome. . . . In the two cases you cite [the *Florida Air* and *Miss Renee*], the stations performed admirably.[9]

Captain Goward's statement is correct, but not forceful enough because he implies that the system has suddenly gone awry. The real problem is that the system has always had inherent flaws that can be traced back to 1915.

Project Kimball will have an executive steering committee, a guidance team, and a project team. The executive steering committee consists of six admirals, two captains, and one master chief petty officer. The guidance team consists of 17 captains, four commanders, one chief warrant officer, two master chief petty officers, and two civilians, with five captains and one civilian as ad hoc members. The 17-member project team has not yet been completely filled, but already has a captain, a commander, a lieutenant commander, and two lieutenants on board. In short, of 42 members chosen to decide the direction of the small-boat stations in the future, only four people have any experience at a small-boat station.

The system currently in place at U.S. Coast Guard headquarters regarding the small-boat stations has obvious flaws that can be easily remedied without additional cost to the service and taxpayers. The system can work if the chain of command listens to the experts—those who serve at the units. Captain Goward has written that "long-term change for the better will require cultural shifts that will be both difficult and painful." Yet it appears that Goward is claiming that the enlisted force must come to grips with the presence of officers at the units. The real cultural shift should be that the U.S. Coast Guard senior leadership must learn to listen to their senior enlisted force. If such listening occurred, there would be little need for organizational changes. Will this listening take place? Probably not, for the senior leadership appears to have no real desire to undertake the required "difficult and painful" cultural shift.[10]

Senior Chief Michael P. Milligan, writing later in the U.S. Naval Institute *Proceedings,* stated, "Captain Goward's suggested intrusion of officers into the small boat community disturbs me. The small boat station does not need a commanding officer. [That is, a commissioned officer.] The senior enlisted officer-in-charge or warrant officer has earned the position with a career's worth of experience, competence, and professionalism. There is no operational or cultural necessity that would require placement of an officer in this mission-focused environment."[11] Senior Chief Milligan continues, "Adding officer billets is not the answer in supporting the small boat community. The Coast Guard already has too many officers and many of the problems within the enlisted community are a result of lack of direct support to the [enlisted] people. While support for officer programs and training education remains a top priority, the service routinely cuts back or ignores enlisted training needs and requirements. This lack of support was identified in a Workforce Cultural Audit—yet another study emanating from headquarters—as a major concern and is still a reason why people are leaving the service."[12]

The comments of Senior Chief Milligan on the service's lack of support of enlisted personnel illustrates one of the more interesting facts about the new U.S. Coast Guard. Years ago, it was usually the crusty old chief petty officers and salty first-class petty officers who would vociferously harangue about officers. In today's U.S. Coast Guard, even seaman apprentices charge that they would not like to have officers at the stations, because they would worry more about their Officer Evaluation Reports than making hard decisions. A captain asserts, however, that seaman apprentices have "zero experience" in making such statements. The seaman apprentices, the captain claims, are just parroting what chief petty officers say. While this assertion may be true, that is not the case for seaman and above. After several years of research on the small-boat stations of the U.S. Coast Guard, it has become apparent to me that officers and enlisted personnel are widely separated. According to one lieutenant commander who has no experience at boat stations, "Chiefs, Senior Chiefs, and Master Chiefs are too likely to tell young Coasties to charge across the bar, whereas junior officers are more proficient at analyzing situations and using risk assessment principles." Other officers have commented that senior enlisted people are not able to do something as elementary as making up a watch list and that this is one of the reasons for the many hours people at a station must work. Perhaps many young coxswains see surfmen and senior petty officers working long hours and getting little respect from commanders, and wonder why anyone would wish to continue in that direction.

Throughout this book I have recorded what many at the units feel are problems at the small-boat stations and how these problems contributed to the deaths of the three U.S. Coast Guardsmen at the Quillayute River station. I have also included observations about the problems and the responses from the leadership. These problems can be broadly summarized as a lack of communication between headquarters and the stations, a poor personnel policy, a shortage of personnel, a poor attitude about the stations by the senior leadership, and a lack of sufficient funding for the units.

Although it is easy to note faults, it is quite another thing to lead an organization wrestling with "the Big Picture." Admiral James M. Loy faces considerable problems in leading the U.S. Coast Guard at the beginning of the 21st century. Furthermore, there are no quick fixes to the problems illustrated in this book.

It would be dishonest not to admit that some of the problems of the stations are exacerbated by the people at the units. Stations have a tradition of poorly documenting their actions, which is unquestionably caused by too few people and too little time. Moreover, few people in the U.S. Coast Guard write, and even fewer at the small-boat stations. Lastly, the senior people at the stations must communicate their needs to headquarters in a responsible, professional manner. Yet what has happened over the years is a figurative and literal shouting match between the two groups, one that has become a way of life, with neither side listening. Both sides must respect and listen to each other to accomplish the search and rescue mission.

One of the problems identified by Captain Goward and others is a lack of qualified people. To have a core of well-qualified people at the small-boat stations, the boatswain's mate rate—the rate that drives the boats and runs most stations—could be specialized into three groups. One group would be assigned only to stations. Because the leadership will never stop sea/shore rotation, this group could be transferred to patrol boats, whose work is much like the tasks at small-boat stations. Another group would rotate between aids to navigation teams (ANTs) and vessels engaged in aids to navigation work. The third group would work at marine safety offices, and sea duty would be aboard medium- and high-endurance cutters. The people in the E-2 and E-3 grades could be detailed between the various groups until they chose the field they wished. This arrangement could also work for machinery technicians, the other rate most often assigned to stations.

Tours of duty at the stations should be lengthened. The leadership will invariably describe how sophisticated the new 47-foot motor lifeboat is compared to previous boats. The boat, in fact, is a technological marvel. It seems self-evident that this same technology needs a more highly trained boat force. This training can be accomplished only by spending extra time at a unit learning the craft. But this basic fact appears to have been ignored by the collective mind of the leadership. In fact, crews at all small-boat stations should have longer tours, which would make boat crews more qualified by becoming better versed in the particular operational traits of the region.

The transfer of personnel in summer needs to be reevaluated. While the policy of moving people in the summer months is a "quality of life" issue, this must be balanced by the large number of untrained people arriving at traditionally the busiest time of the search-and-rescue year. No station should ever be put in the situation of Quillayute River in February 1997, where at least 80 percent of the station's personnel had one year or less of experience at the unit. This is a safety issue that applies both to the crews of the stations and to the public they serve.

Many of the possible solutions to the problems of the stations will not come about unless the leadership solves two basic issues. The first deals with the definition and mission of the U.S. Coast Guard. The service has entirely too many missions. In an age that appears set upon less government and less budget for departments, the service must not squander its resources. The basic mission of the U.S. Coast Guard should be search and rescue. It is this mission that separates it from the other four armed forces. This mission is closely followed, in order of importance, by military readiness, aids to navigation, and law enforcement. The largest portion of the service's budget should reflect the basic mission, with remaining funds for the other three missions. If the legislative and executive branches wish the service to perform other missions, then the commandant should insist upon resources before undertaking the duties. If there are reductions in funding for any duties beside the four main duties, then that particular mission should be curtailed.

The best solution for a small organization such as the U.S. Coast Guard is to have everyone enter the service in the enlisted force. After a period of service—perhaps four years—those who wish to become officers would be sent to the equivalent of Officer Candidate School. This would ensure that officers would have a grasp of the duties, in the jargon of the service, "from the deck plates upward." It would eliminate the present situation in which no one in the senior leadership has experience in shore-based search and rescue.

One of the basic problems with Captain Goward's plan to put officers at the stations concerns what officers should be required to know in their new positions. At present, a commissioned officer serving at a small-boat station does not have to be proficient in the boats at the unit. Enlisted personnel who are officers-in-charge, on the other hand, are required to show mastery of the craft. The current plan for the officers at the stations is to have the officers work primarily in administration. As one officer put it, "Officers shouldn't be boatswain's mates." Yet, even if the officers are there for administrative purposes, they are still in command. An academy officer graduating from New London, Connecticut, reporting aboard a cutter as a deck watch officer must learn to become an under way watch officer. Should not an officer assigned to a station also learn how to run the craft he or she is commanding? If officers are to be assigned to stations, they should have the same boat crew training and be at least coxswain qualified.

The U.S. Coast Guard should reevaluate the number of large cutters in its fleet. Perhaps 110-foot patrol cutters and larger buoy tenders could take care of coastal drug interdiction and aids to navigation needs. The greatest assistance the U.S. Coast Guard provided in combat in the Vietnam War was in interdiction work, with the use of 82-foot patrol craft in the Vietnam War. (Larger cutters and other U.S. Coast Guard units also served in Vietnam.) With fewer larger cutters there could be more funds for the stations.

One of the simplest of the remedies to the problems at the stations is also the one that will take the most effort. The U.S. Coast Guard's officer corps has always looked askance at their own service's small-boat community. To them, sea duty is the only true place for someone in the service, unless, of course, there are those important staff jobs ashore that must be filled by officers. Many seemed ashamed to be in the same organization that uses small boats. These same officers do not realize that during World War II the U.S. Navy did ask for help from the U.S. Coast Guard. The request was not for ship drivers, as the navy had plenty of those. Instead, the navy requested people who knew how to operate small boats in surf, those who could operate a landing craft. An interesting thing also happened during the Vietnam War. The navy again needed small patrol boats to help in the interdiction of arms and matériel from the sea. The navy found they had "no such vessels" readily available.[13] Again, the navy turned to the U.S. Coast Guard and their 82-foot patrol boats, whose normal crews consisted of eight enlisted people and a master chief boatswain's mate in charge; a master chief boatswain's mate would have nearly 20 years or more of experience in search and rescue and in boarding vessels, and probably would be in his middle or late thirties. During Vietnam, the leadership of the U.S. Coast Guard decided that each 82-footer needed a lieutenant as the commanding officer and a lieutenant (junior grade) as the executive officer. The reasoning: "The feeling was that the presence of an officer was needed for the job of stopping and boarding vessels. We thought there should be a little more seniority."[14] These officers would be in their mid-twenties, with perhaps one to three years at sea as a deck watch officer. (A U.S. Coast Guard lieutenant and a lieutenant junior grade are the same grades as a captain and first lieutenant in the army and marine corps, or as a company commander and a platoon leader.)

The U.S. Coast Guard's senior leadership must learn that the small-boat stations, performing most of the search and rescue work in the service, justify their existence. Moreover, the stations are crewed by enlisted people who are professionals and who do a professional job.

. . . 21 Reflections

At the beginning of the 21st century, the small-boat stations are still at risk, more so than at any other time since the 19th century. The beginning of the new millennium finds high employment within the United States and a populace wanting less government. (High unemployment usually equals high enlistment rates in the armed forces.) After years of chasing after any mission that would bring more funds to the organization, and more jobs for its officer corps, the U.S. Coast Guard leadership, like the other four military services, faces a difficult time meeting recruiting quotas and slashed budgets. With so many diverse missions within its purview, and an aging cutter fleet, the leadership must now make some very difficult decisions as to just how the service will function in the years to come. The only program within the service that is almost entirely crewed by enlisted personnel is the small-boat stations. Throughout the years since 1915, the stations have been neglected, so that just at this time of slashed budgets and cutbacks the Quillayute River deaths, followed by deaths in the *Morning Dew* case, brought unwanted headlines and critical stories. This notoriety made an easy excuse for cuts: the stations are making mistakes, so they must be unprofessional and should be eliminated. Then suddenly, after 85 years, the officer corps "discovers" that station personnel work very long hours and can, on occasion, make wrong decisions. With the surprise of a person receiving an electrical shock, it became apparent that the equipment at the stations was dated—or ancient—and that boat personnel experience is plunging. Officers decry the fact that the stations have no official doctrine. The answer so far is a plan to perhaps put officers at the Groups and stations. But, as one surfman exclaimed: "We need more E-3s [seamen], not O-3s [lieutenants]."

Many of the decisions made in the late 1990s by headquarters came about because we live in a litigious society. The headquarters staff of the U.S. Coast Guard rightfully must be wary of the chance for claims against the government. This can also be the reason for an undercurrent of feeling that if one can successfully pass enough regulations, one can eliminate all the risks in small-boat operations during bad weather: thus, no deaths or litigation. When this wariness is combined with an officer corps that cannot make a mistake because mistakes will reflect badly on their Officer Evaluation Reports and stop promotion, then some of the actions taken by the leadership are a little more understandable, though not necessarily correct.

Because the U.S. Coast Guard does have too many missions, along with an aging inventory, it appears the service has made a decision about which way it will proceed; "Deepwater," the project to design a new cutter fleet, has priority. This priority should not be too surprising, as officers are in command of cutters. The announcements that the stations in some groups in the Thirteenth Coast Guard District now have to make about not responding appear to be the first step in a move to excise this enlisted program. On the current course, the small-boat stations may disappear at the end of the service life of the 47-foot motor lifeboat, except in politically sensitive areas such as Massachusetts.

Can the U.S. Coast Guard still support the small-boat stations in the way they should be supported? The response to this question is a firm yes. Apparently Admiral Loy, through Admiral Riutta, wishes the public to recognize that the U.S. Coast Guard does search and rescue, for he stated that all units in the service have search and rescue as a primary mission. Even though that statement can be questioned, the 188 small-boat stations save most of the lives and property credited to the U.S. Coast Guard each year. Perhaps the problem within the U.S. Coast

Guard is that it has not yet acknowledged that what makes the service unique is its search and rescue. None of the other services has search and rescue as a main mission. To be sure, all of the services provide search and rescue to help recover their own personnel from a mission, such as from a bombing attack. The other four military services have also provided search and rescue services to the civilian community when requested, but only the U.S. Coast Guard provides rescue on a regular basis to both the civilian and military communities. Once again, the service's own statistics show that the small-boat stations do the great majority of this work, and the stations are an enlisted program.[1]

The answer to the above question is at the crux of the issue. The proliferation of the many missions now undertaken by the U.S. Coast Guard seems to have been aimed at increasing the number of billets for officers. Again, if one accepts that search and rescue in the service is one of—if not *the*—most important missions, then there needs to be a severe reduction in the number of the academy officer corps, a reduction that in reality will not happen. To keep both a strong officer corps and a strong small-boat station search and rescue force, the U.S. Coast Guard needs to recognize the worth of the stations, fund and staff them adequately, and limit the amount of additional missions it accepts. This may require the reduction of a number of cutters, and even some stations, but it would make for a stronger and more viable service.

One of the troubling aspects of the Quillayute River deaths is how the people in the service perceived the investigation into the deaths. Many of the questions that are covered in this book are questions asked in the field. To say that there is bitterness even after three years is to understate on a grand scale. One person angrily exclaimed: "I thought that was a major shit trick that D13 [Thirteenth Coast Guard District] pulled on that poor bastard [Bosley]—those sons a bitches! I'm still hot about that. I didn't like Dave, and to be honest with you, there were times I would have loved to kick him in the short hairs myself, but I sure as hell would have never pissed on his grave like those miserable bastards did."

Chief Warrant Officer F. Scott Clendenin remarked,

> What bothered me the most is that the motor lifeboat community saw one of our own perish, and we saw people coming into our lifeboat community and pointing fingers. These young, junior people are going to make mistakes and the senior people have to be there to cushion the blow when it hits. There was no cushion for Bosley or the crew. There was instant finger pointing.
>
> Did Bosley make a mistake? We will never know. We do know an accident occurred. Did we go into this thing wanting to know how to do the job better, or did we go in looking for blame? Unfortunately, we went in lookin' for blame.
>
> Did the surfbelts fail because it didn't have a certain type of clip on it? No, it failed because we put that motor lifeboat into the extreme situation, because the boat did roll and it came up with everybody on it. The boat kept running all through the whole accident. But we did not learn anything, we went in and pointed a finger and said, "It was him! It was him!"
>
> In the aviation community, they put a helicopter in the drink and they came out with something. Hey, when the seas are running you gotta set that auto leveler higher than the 15 [feet] programmed. They learned something. A junior pilot learned from a senior pilot on a safety standdown from that aviation incident. You never heard that the pilot or crew failed; instead we learned [to set the auto leveler higher]. Unfortunately, when you come in right off the bat with that finger blaring, people are going to shut down and not say things that they need to be saying.

The Coast Guard's sudden interest in the stations has produced suspicion within the ranks, and there is good reason for this wariness. Headquarters has a track record of 85 years of inattention and numerous studies that admit their own neglect. So, what has produced the sudden interest? Two reasons stand out.

The service needs to revitalize their cutter and aircraft fleet. The service has finally come to the realization that most Americans know very little about the U.S. Coast Guard—due largely to its own unwritten policies—but what the taxpayers do realize is that small white boats save people. As such, the service is a humanitarian organization that rescues those in distress. In order to obtain the necessary funding for cutters and aircraft, under the project name of "Deepwater," the service must show that it cares for the stations. In other words, the leadership has decided to "ride on the backs of the stations" to obtain funding. If the Deepwater project is successful, perhaps the interest in the stations will again quickly wane; as everyone within the leadership knows, "small boats, small problems."

The other reason for the stirring of interest after 85 years of inattention by headquarters is a desire to bring an end to a tradition that has been in effect since at least 1878. The stations have fended for themselves, except for periodic inspections and sudden short spasms of interest, usually after deaths, which has produced a fiercely independent, enlisted professional lifesaving force. This force has truly lived up to the service's motto: *Semper Paratus*, "Always Ready." The enlisted crews have worked long hours, with very little help from an officer corps that seems to give only reluctant lip service to the motto, especially when it comes to shore-based lifesaving. After 85 years, the service has awakened to the fact that they have developed within their midst a professional group of enlisted people that does not really follow the bureaucracy of the officer-dominated organization. This is tantamount to heresy: This cadre of lifesavers needs to be brought under control. With very little understanding of what they are trying to control, the service is about to set into motion something that could spell the end of a noble tradition. Recent pronouncements have indicated that the service feels the stations have been a little too "always ready," that perhaps they should "back off" a little from their readiness posture. This, of course, will allow the crews to fit into the new "fatigue standards" without having to give the stations the correct number of people at the units. Some at the stations wonder if the new motto should be "semi-ready."

Some of the senior people in the enlisted ranks at the stations have indicated that much of the major change taking place is of their own doing. "We have yelled, screamed, and pleaded for years that no one is listening to us," these senior people admit. "So, we 'sucked it up' and went about our jobs the best we could. They finally start to listen, and look what they are doing. We should do just enough of the new way to keep them away from us, suck it up again, and get back to saving lives the right way." Many senior enlisted people, in short, believe that after the latest spurt of attention and once new bureaucratic controls are placed upon the units, they will again be forgotten.

Some officers have responded to my comments about the officer corps and the stations. One officer claimed that there are more officers now in the U.S. Coast Guard because "the increased complexity of our technical and regulatory activity has created a bunch of jobs requiring backgrounds that can't be rented for less than officer pay." This is a reasoned approach and has merit. During the research for this book, an officer maintained that he was "concerned that I might harm the gains that have been made" in the small-boat community. Many years ago this type of statement was another way of saying, "Don't rock the boat; let us take care of the problem." Lastly, there are many both within headquarters and in the field who think writers never present the facts about the service correctly, especially in cases where deaths have occurred. This writer encountered numerous difficulties in reviewing "the facts" for this book; notice how difficult the Coast Guard makes the process.[2] The odyssey of the request to Admiral Loy, described in chapter 17, is a good example. All of this suggests two things: People working at the small-boat stations consistently say that headquarters does not listen to them. If districts and headquarters will not respond to a researcher who is trying to tell the story of the stations, then it appears they can also be indifferent to the units. Secondly, headquarters apparently does nothing to help the public receive the "facts" in a timely manner.

Other officers have argued that I should not use "second and third hand information" gathered from the stations—that I will learn the "facts" only by seeking materials from headquarters. Yet who has the most experience, officers in headquarters or those have worked in the field for over 20 years?

Americans should realize that during all of the maneuverings described in this book—and many more that are not described—the men and women at the U.S. Coast Guard's small-boat stations have continued to make the stations ready in the best manner they can for the time when, some gale-swept night, the word "Mayday" comes screaming and crackling over a radio's loudspeaker. They have accomplished their work despite little assistance from their own service and very little acknowledgement from the people they serve. CWO2 F. Scott Clendenin described it the best: the small-boat community of the U.S. Coast Guard "is a world of so much passion and intensity to do the right thing." It is unfortunate that these men and women have been largely overlooked while doing a dangerous job. It would be an affront to all those who have gone into harm's way to help strangers in distress if the small-boat stations of the U.S. Coast Guard disappeared.

The events at Quillayute River remain indelible in the minds of those closely associated with them. Most of the people involved or close to the deaths are still marked by the early morning hours of 12 February 1997. While being interviewed, some stopped for a minute, trying to gather their thoughts; when they continued, their voices strained a bit but they carried on. Others tried to cover their emotions through a gruff exterior. One person broke down and cried.

Some of those who were at Quillayute River noticed that when they were transferred to other units, their shipmates looked at them wondering and eventually worked up the nerve to ask about that night in February. Seaman John Stoudenmire said it best: "Everyone had to deal with their own demons."

Ben Wingo thought that people looked at him questioningly, wondering whether the event had affected him mentally. While attending a service school, Ben handled this in his typical way: When classmates asked Ben if he had seen the survivor of the Quillayute River incident, not knowing it was Ben, he would reply, "Yeah, he's all screwed up. He, like, cowers in the shower."[3]

"I couldn't believe it, actually," said Jeremy Gustafson, who's stationed with Wingo. "He's so different from what I would have expected. I would have been devastated, and he's so happy."[4]

Ben says that sometimes when he thinks of that night, he changes the ending. He imagines they snatch a couple of travelers from their sinking sailboat and return to the rescue station victorious. He imagines everyone survives. "'I wish we could just turn back time. That would have been great to have had a . . . rescue."[5]

"Don't make me into a saint. I'm not," Ben has said. "I'm just lucky. I lived, and they didn't and there's really no reason why. That's just the way it happened."[6]

Jon Placido gives Ben Wingo much credit for sticking at the station after the deaths. He could very easily have requested, and received, a transfer. As Placido put it, "If Wingo could take it, no one else at the station could complain."

Captain Philip C. Volk, the former commander of Group Port Angeles, said:

> I will remember the day, the night, the week for the rest of my life. Four young kids . . . went out in the teeth of a gale and three [died]. And that is what we do every day. I have had a lot of people say: 'My God—I had no idea what you did. . . . These guys go out and are killed. What you guys do is real.'
>
> If the guardian angels of those people on our boat went somewhere else, they landed on Paul, Ray, Neal, and Chuck's shoulders and helped out saving those people from the *Gale Runner.*[7]

Captain Paul A. Langlois, pilot in command of helicopter 6589 from Air Station Port Angeles, commented, "We all believed something miraculous helped us. Somebody above was watching over us. There must have been enough pain in losing our own crew that somehow it was right to help those two people."

Commander Raymond J. Miller, the other pilot in helicopter 6589, would later remark that the battle by The Needles gives "a heightened sense of being connected to the process. In a way it comes from fear and fueled by adrenaline. It is a very invigorating experience, too. Not that you would ever want to buy a ticket and do it again, not on purpose anyway. There is that sense at the end of it of gratification, not pride, but, my goodness! look at what people and machines can do when they really have to."

Tom Byrd, now retired, who was the boat engineer on the second boat, said, "I still think about the crew on the other boat. Matt was my best friend at the station. We used to sit on the couch all the time and shoot the breeze." Tom represented the station at the funeral in Missouri. After his retirement, Tom stopped off on his way home to see Matthew Schlimme's family. "Matt's family basically calls me their second son. It was really hard going to see Matt's grave. I remember looking at his tombstone. It was really hard doing that."

There is another side to the aftermath of the Quillayute River deaths that is not pleasant. When I was nearing the completion of this book, BM2 W. Brent Cookingham called. On that tragic night in February 1997, Surfman Cookingham had been off duty and was recalled back to the unit. Brent ran the station in a very professional manner when Master Chief LaForge went to First Beach upon learning that Miniken had washed up on the beach. After four years, I can still see Cookingham working with the watchstanders in the communications room and speaking in a calm voice. Brent had a desire to serve aboard the cutter *Eagle,* the U.S. Coast Guard Academy's sailing vessel. Nearing the end of his tour at Quillayute River, Brent wanted some time away from the stations. The only thing the detailer would give him was a 270-foot cutter out of the Norfolk, Virginia, area. Brent instead opted to take another tour at a station, as the stations were still short of surfmen. About this time, a request came through for boatswain's mates for temporary duty aboard the *Eagle.* Brent checked with the people from the *Eagle* and found they were anxious to have someone of his experience. Master Chief LaForge worked with the detailer to have Brent transferred to the *Eagle* and then to his new duty station at Morro Bay, California. At some time in this process, someone either at headquarters or the district did not like the arrangement and stopped the move. As a result, Brent ended up going to Morro Bay without serving on the *Eagle.*

During our conversation, Brent remarked, "I have some news that you might as well hear from the horse's mouth. About four months ago, I started experiencing nightmares and suffering from insomnia. I have been diagnosed with PTSD [posttraumatic stress syndrome], with severe depression and severe anxiety. I am no longer allowed on the boats. What they think might have kicked it off were three very bad cases almost in a row at this station. One dealt with me on the 47-footer going into shallow water with 18-to-20-foot breaks after body surfers and a park ranger who had tried to help them." Brent went on to say that the 47-footer developed overheating problems so that he could not make it back into Morro Bay in daylight and arrived off the bar at night. As the station did not have a backup boat, he then had to proceed in very heavy seas to another area to seek shelter. "The Coast Guard was going to give me a medical discharge, but I was able to get help and I will now retire in December." Brent is the second crewman who has admitted to being under medical care because of 11–12 February 1997.

More than one person has remarked that they know Master Chief LaForge and BM1 Placido believe they are responsible for the deaths of the three U.S. Coast Guardsmen. This account demonstrates the fallacy of this thinking. What has not helped is that more than one person has admitted that new recruits at the U.S. Coast Guard's boot camp at Cape May, New Jersey—after being ordered to Quillayute River—were told: "You are going to a station that is screwed up and kills people."

The incident at Quillayute River is a perfect example of the work accomplished by the small-boat stations of the U.S. Coast Guard since the shouts of "ship ashore!" were first screamed along the coastline of the United States. It illustrates the extreme danger faced by those in very small craft against a very large, furious, and unforgiving sea. By examining the incident, we are able to see the training and glimpse the lives of the people who serve at the stations along the coastline of this country.

The incident also shows the frustrations felt by those who do push out into storms to help those in peril upon the seas.

It is only natural for someone looking at a gale-swept, towering sea, whipped to a white froth by the wind and making a sound like an express train when the swells are approaching in the dark, not to want to go out into the maelstrom. Common sense dictates that the people aboard a boat less than 52 feet in length in such conditions are in for a hard, dangerous time. Chief Warrant Officer Clendenin has said, "Anyone who goes out into such conditions, when all your senses are screaming not to go, is a hero."

Charles A. Lindenmuth II, the emergency medical technician with the Forks, Washington, ambulance corps, best summed it all up: "Until Wednesday, February 12th, at three o'clock in the morning I had taken these people for granted. When we were finishing working with them on other ambulance runs we turned around and went back to our homes and families. On the other hand, the Coast Guard personnel got to go back to their efficient, modern, but lonely station and military housing. I will never take these courageous people for granted again. Nor will I ever forget what I saw that stormy night, or how I felt the first time I realized these brave people put their lives at risk every time they effect a rescue. They did not lose their lives. They voluntarily and unselfishly give their lives, in the true tradition of the United States Coast Guard, that those in distress might live. Every rescue attempt by members of the United States Coast Guard carries with it a degree of risk, yet these brave young people continue to perform rescues whenever they are summoned; battling the weather, the seas, and the odds."[8]

On 12 January 1961, on the Columbia River, the U.S. Coast Guard suffered one of the single greatest losses of people from one of its small-boat stations since 1915. Five U.S. Coast Guardsmen were killed going to the assistance of the fishing vessel *Mermaid.* In 1982, BMCS Darrell Murray, U.S. Coast Guard (Retired), a participant in the case, attempted to obtain from the U.S. Coast Guard transcripts of the hearing held on the incident for a claim to the Veteran's Administration. Murray was thwarted in his attempts. A letter to his U.S. senator brought the following response from U.S. Coast Guard headquarters: "We have conducted a thorough search of our files of such investigations, both at Coast Guard Headquarters and the Thirteenth Coast Guard District Office in Seattle, Washington. Although every possible subject category was searched, we have, unfortunately, been unable to find any record related to the incident to which Mr. Murray referred." In short, even though the incident was known among the people serving in the Columbia River area, officially the case never happened. Senior Chief Murray, on his own, eventually found another location to search and finally, on 10 March 1983, headquarters was able to locate the report. In only 21 years, U.S. Coast Guard headquarters forgot a major loss of life at one of its small-boat stations, although the loss continued to be remembered at the local level.[9]

The purpose of this book has been threefold: first, to show the work of the men and women of the U.S. Coast Guard's small-boat stations; second, to show the great differences between those who conduct search and rescue from the stations and those who are far removed from the scene but who are responsible for the policies affecting the units and the boating public. Third, and most important, this book is an attempt to ensure that all those who participated in the rescue attempt of the sailing vessel *Gale Runner* are not forgotten as quickly as those who worked the *Mermaid* case.

Epilogue

On Saturday, 12 February 2000, I again made the long drive to the Quillayute River station. Since that tragic night in 1997, the drive has not gotten shorter, nor easier. This was the first time in three years, however, that the weather cooperated: the sun actually shone for a good part of the day, a rare occurrence for February. On the first anniversary, the weather was as bad as that night in 1997. The second-year anniversary brought a bar so rough all one could see was high white foam.

As always, the people at the station greeted me as if I were a member of the crew. (They have not asked me to stand watch, however, for which I am grateful, as are the officer-in-charge and the XPO.) There is now no one at the unit who served there on 12 February 1997. BM1 Michael "Mike" Saindon replaced BM1 Jonathan "Jon" Placido as XPO in 1998. Mike is also a surfman and fills one of the surfmen billets—the station is still short one surfman. Mike came in from home to talk to me. He has a very strong interest in U.S. Coast Guard history, and we discussed this and other matters.

Mike Saindon entered the U.S. Coast Guard in 1988 "to drive boats in the surf and for the search and rescue. It's funny, but driving boats in the surf is only about 5 to 10 percent of what I do." After boot camp, Mike served at the station/base Southwest Harbor, Maine, for two years, reaching the rate of BM3 before a transfer to the 378-foot cutter

Sherman (WHEC-720) at Alameda, California, for two years and making BM2. He then transferred to station Depoe Bay, Oregon, for five years, where he was promoted to BM1 and made surfman. His next assignment took him to the 82-foot patrol boat *Point Stuart* (WPB-82358) at Corona del Mar, California, as XPO for two years before coming to Quillayute River as XPO. He will make Chief Boatswain's Mate shortly. Mike has taken a one-year extension at the unit.

BMCS David "Dave" Meyrick, another surfman, took over as officer-in-charge from BMCM George A. LaForge upon his retirement in 1999. Dave entered the U.S. Coast Guard in August 1980, and his first duty station after boot camp was in the patrol boat *Point Countess* (WPB-82335) out of Port Angeles, Washington. Dave's next assignment was to the Aids to Navigation Team at Port Angeles. While stationed in Port Angeles, Dave met his wife, Alma. After the Port Angeles assignments, he moved to the U.S. Coast Guard Air Station in Kodiak, Alaska. Then Dave transferred to the motor lifeboat station at Umpqua River, in Winchester, Oregon. Here, Dave made surfman. He next reported to the patrol boat *Point Ledge* (WPB-82334) at Fort Bragg, California. Promoted to chief petty officer, Dave served as XPO of the motor lifeboat station Chetco River, Oregon. After a three-year assignment, he became officer-in-charge at the station in Maui, Hawaii. I mentioned that Quillayute River must be a major cultural change for him. Dave responded that he was "glad to get back to the rain again and did not miss Hawaii at all."

Because the weather was so nice on this third anniversary, some of the crew took the small rigid hull inflatable boat out for a familiarization trip to the north and also to give it a test run after some recent repairs. One of the crewmen aboard was MK2 Bryn Szito. The rest of the deck force was working on one of the 44-foot motor lifeboats. A typical day, even if it was a Saturday: training and maintenance.

One of the interesting things about visiting the small-boat rescue stations of the U.S. Coast Guard and taking the time to talk to the crew is learning about the talents and interests of the individuals who make up the station's complement. I often recall the interests and talents of the 1997 crew. BMCM George A. LaForge is a sailing enthusiast. One of the more ironic aspects of the incident is that a sailboat caused so much trouble for George. BM1 Jon Placido loved to duck hunt.

One of the things that I remember is saying to BM2 W. Brent Cookingham, "I will always remember how the second boat looked that night when it hit the first big wave. Sometimes I wish I could paint."

Cookingham looked at me and said simply, "I paint."

I had an empty file folder in my hand and I sketched on it, in a very simple and primitive manner, what I saw and how I thought such a painting should look. Brent said he would do it, but as he was busy it would take some time. Less than a month later, however, I received a call from Brent saying, "I got inspired. It's finished. You want to come out and look at it?" Brent's talent produced the best depiction I have seen of a motor lifeboat entering high seas at night. The painting now hangs near my desk and is the dust jacket illustration for this book.

I also recall Ben Wingo's abilities as a soccer player and MK2 Tom Byrd's interest in the history of World War II in the Pacific. BM2 Donald Miterko liked riding bicycles. I remember the voice of the quiet-spoken John DeMello as we talked about the birds in the area and about Hawaii.

I also vividly remember all of these people with their drawn, haggard faces on 12 February 1997. Of the three who did not come back, I still see Matthew E. Schlimme's smile on the night of 11 February.

The present crew of the Quillayute River station is as diverse and interesting as those men and women of three years ago. Mike Saindon enjoys reading about U.S. Coast Guard history and collecting books on the subject. SN Tylor Foster wants to write and likes to read and discuss writers of American literature such as F. Scott Fitzgerald and Ernest Hemingway.

DC2 Kevin Carr loves fishing, guns, and his yellow lab, Rusty. His wife, Carol, moved back to Florida because she had problems with the isolation and some people in the tribe. Kevin will be transferred to a station in Florida this summer, and they will be able to pick up where they left off.

SN Jennifer Gonzales, from California, loves to shop. She drives at least five hours to Seattle on her time off, just to walk the malls. If she does not do that, she likes to see high school games just to be in a social setting.

BM3 Tim Tregoning loves racing boats. He spends his off-duty time rebuilding engines and working on his boat.

During the day of my visit, the father of Clinton Miniken visited the station, but did not come inside. Later, BMCM George A. LaForge and his wife, Melva, visited. We did not talk much about why we were at the station.

At lunchtime, I saw some clowning around between two crew members; one was carried into the scullery and drenched in water. The other said, "That makes us even." Chow time at stations has always been the time for a crew to come together. At good stations, it can be a pleasant time for everyone. At bad stations, people gulp their food and get away as soon as possible. At the Quillayute River station, despite their isolation, the meals I have shared at the unit have always been pleasant times.

After lunch, I continued to talk to Mike Saindon. We were interrupted by the communications watchstander. MK2 Szito's wife, Stephanie, who was pregnant, had developed pains and needed to get to the doctor. Mike called his wife, Susie, and asked her to go over to the Szito's quarters to watch their children while Stephanie was driven to the doctor. Mike then had the rigid hull inflatable boat recalled so that Bryn could be with his wife. This is not the first time I have seen the executive petty officers and the officers-in-charge at stations have to handle emergency domestic matters.

Three years have brought change to the station, but it is a type of change that never really changes. New people arrive, are trained, and then depart. I continue to marvel at some of the edicts the station receives and how, despite some strange rules, the crews continue to be willing to jump into their motor lifeboats to help strangers. This willingness comes from the dedication of the officer-in-charge and his executive petty officer, and from the crew's innate sense of duty, despite some questionable leadership from higher up the chain of command and lack of recognition by those they serve. The unforgiving ocean remains the same.

Fig. E.1. The memorial to those who died on CG 44363. Photo by Dennis L. Noble.

For the last three years, I have continually voiced the comment that the entire crew of the Quillayute River station never received the credit due them on 11–12 February 1997. A few members among them received awards, but not the station as a whole. Perhaps there is no such award, but there should be.

I had brought the tapes of the radio transmissions of 12 February for Mike to listen to and talk about with his coxswains. Mike asked what was happening at the time of some of the transmissions. I quickly found myself uncomfortable and realized that even after three years it bothered me to listen to the broadcasts. I decided to leave earlier than planned.

Once outside, I stopped to look over the scene from the front steps where I had rushed out in the early morning hours of 12 February 1997 on the way to the bar with Master Chief LaForge, hearing the static-filled "we rolled the boat." Today, with the sun shining, James Island and the small islands to the right dominate the surroundings.

There is one major change on the outside of the station since February 1997. Looking seaward from the station's steps, to the right and up a short walkway, one sees a monument depicting a motor lifeboat in a breaker (fig. E.1). The monument was bought by funds collected by the

Port Angeles chapter of the U.S. Coast Guard Chief Petty Officer's Association. It is one of the finest memorials that I have seen in this country to enlisted people from a station who lost their lives trying to help others.

I walked out to the monument. Roses lay at its base, put there by Clinton Miniken's father. The main inscription, from a report written by a U.S. Revenue Cutter Service officer on a U.S. Life-Saving Service rescue in 1885, says,

> These poor, plain men, dwellers upon the lonely sands, took their lives in their hands, and at the most imminent risk, crossed the most tumultuous seas . . . and all for what? That others might live to see home and friends.

I departed for home.

Postscript

Just as the manuscript for this book was nearing completion, two events happened that have a bearing on this story of the deaths at the U.S. Coast Guard station at Quillayute River. One event deals with yet more deaths in the small-boat community, this time from station Niagara on Lake Ontario, and the other concerns the Quillayute River station itself.

The only news so far released by the U.S. Coast Guard on the recent deaths is that four Coast Guardsmen were under way at 7:45 P.M. on Friday, 23 March 2001, "conducting a routine law enforcement patrol and area familiarization . . . [when their] 22-foot boat" was hit by a wave and capsized, throwing the crew into the frigid waters of Lake Ontario. When station Niagara was unable to contact the boat by radio, a search began. It was not until approximately 12:30 Saturday morning that the crew was pulled from the water. BM2 Scott J. Chism, 25, and SN Chris Ferreby, 23, died.[1] Unlike the Quillayute River investigation, the results of the initial investigation have not been released quickly, although there was an announcement to all stations to ensure people had the proper training in survival equipment, especially exposure gear.

The 6 March 2001 issue of the Port Angeles, Washington, *Peninsula Daily News* announced to its readers that the "U.S. Coast Guard Station Quillayute River at LaPush is being considered for closure." The newspaper quoted the surface operations officer for Group Port Angeles, Lieutenant Commander Fred Myers, as saying, "Based upon historical data there isn't enough to keep that station in its present form." The initial story went on to quote Myers as saying the station could "close, get smaller or nothing" could happen. Myers said, "There's many places along the coast where the Coast Guard isn't, there's always a risk." Myers went on to say there would be public hearings on the subject.[2]

The public affairs office of the Thirteenth Coast Guard District dubs the proposed closure of the station a part of the "Olympic Peninsula Service Improvement Proposal." In news releases, the district's reasoning for the change of status centers on population changes, search and rescue workload, and possibility of accidents in areas not covered by U.S. Coast Guard units. According to the public affairs office, "The general population in the Puget Sound region and the number of registered boats is increasing annually. These increases along with the need to meet drug smuggling activity, enforce laws for commercial and recreational fishing, and the scaling back of state and local agency services . . . prompted the action." Furthermore, the district points out that there is a location in the Puget Sound region where the Coast Guard has "no surface response asset" close to where a Washington State ferry crosses "commercial shipping lanes 18 hours a day." In the meanwhile, according to the district, the search and rescue workload for the Quillayute River station has decreased over the past five years. According to planners, the station "has responded to an annual average of less than 28 cases in the last four years, and has only responded to six cases where lives were at risk since 1993." To remedy the problems, the district proposes to move the people and boats attached to the Quillayute River station "to other areas in Washington State where the demand for Coast Guard services is increasing."[3]

The Quileute tribal council was quoted as being "opposed to losing protection for its fishing fleet and a non-Indian fleet based in LaPush." Reaction to the news in Forks centered mainly on economic factors. Diane Schostak, director of the Forks, Washington, chamber of commerce said the closure "would affect the area's efforts to promote water-related activities, such as kayaking and fishing tours." Schostak also stated that "Forks and LaPush businesses would lose about $1 million annually from the station's 26 active duty personnel if they were transferred—a loss Forks can't afford."[4]

Clallam County Commissioner Mike Doherty, who represents the West End on the County Commission, also noted that the closure would hurt the economy, and that it could "affect marine wildlife protection as well." Doherty pointed out that if an oil spill occurred in the nearby Olympic Coast National Marine Sanctuary, the lack of station personnel would prevent spill crews from quickly surveying damage. "The marine resources deserve protection," the commissioner said.

Doherty also questioned the accuracy of the statistics the coast guard officials were using for their decision. Thirteenth Coast Guard District officials claimed that other U.S. Coast Guard units in the area received more cases than the station at LaPush. The commissioner, however, noted that the "LaPush calls are generally more serious than the often minor or false alarm calls elsewhere."

"In my mind, the severity of the calls is much different," said the commissioner. "It doesn't make much sense to increase use, then decrease protection for the users."[5]

Adding to the mix, the *Peninsula Daily News* ran an article in their 9 March issue about a $91 million shortfall in the U.S. Coast Guard's budget. Thereafter the paper, not too surprisingly, began to link the closure of the Quillayute River station with the budget crisis within the service.

In the same issue, a page one headline announced that the fate of the station could be known as early as June. Yet CWO Chris Haley, public affairs officer for the U.S. Coast Guard, was quoted as saying, "We have not made a final decision yet. . . . At this point, we really want to enforce the idea we're here to listen to concerns."

Forks attorney Rod Fleck told a different story. Fleck said the "Coast Guard officers who briefed city officials on the possible closure . . . presented information leading staff to anticipate an early decision." Fleck's observation seemed correct, for by 27 March a page one headline stated: "Little hope given for LaPush site." The article stated, "Public opposition alone may not save the U.S. Coast Guard's Quillayute River Station . . . from closure. . . . Barring political pressure and unforeseen facts raised at public hearings . . . the station will likely be shuttered, Chris Haley, a spokesman for the 13th Coast Guard District, said."[6]

An interesting dichotomy has developed between what is being reported in the newspaper and what those in the Thirteenth Coast Guard District say they are saying. When asked to comment on the *Daily News*'s linking of the station closure to budget deficits, CWO Haley replied, "The current proposal is not tied to the current budget shortfalls. The paper keeps making the connection despite our corrections. Budget is an issue as we must look at the current state of resources and how we can meet the risks with what we have, but the budget is not driving the proposal. We . . . look at our resources, outcomes and risks on a regular basis. This is a part of that on-going look."[7]

In response to why one coast guard official seemed to be saying the public hearings did matter, while another said they did not, Haley responded, "The meetings were held to ensure that we looked at all of the necessary information to make an informed decision. No decision has been made yet. We still have a lot of work to do. This is still just the data collection phase."[8]

When asked for the reason for the current U.S. Coast Guard budget shortfall, Haley wrote, "The current budget shortfall has been attributed to rising energy costs and an increase in both payroll and health costs. The payroll and health costs were mandated by Congress without any additional funding."[9]

As the manuscript for this book was sent to the publisher, the investigation into the deaths at Station Niagara has not been released and the decision on the closure of the Quillayute River station has not been reached.

Appendix 1

The Main People in This Story and What Happened to Them after 12 February 1997

ADM Robert Kramek—Commandant of the U.S. Coast Guard at the time of the deaths at the Quillayute River Station. Retired in 1998.

VADM James M. Loy—Appointed Commandant of the U.S. Coast Guard in 1998 and promoted to admiral.

RADM J. David Spade—Retired from the U.S. Coast Guard and lives in the Pacific Northwest.

CAPT Philip C. Volk—Retired from the U.S. Coast Guard and lives in the Pacific Northwest.

CDR Paul A. Langlois—Transferred to U.S. Coast Guard Air Station Savannah, Georgia, as commanding officer. Promoted to captain. Now serving at U.S. Coast Guard Personnel Command, Washington, D.C.

CDR Raymond J. Miller—Transferred and is commanding officer of U.S. Coast Guard Air Station Savannah, Georgia.

CDR Michael A. Neussl—Transferred from Astoria, Oregon, to U.S. Coast Guard Air Station San Francisco, California, as commanding officer.

CDR James M. Hasselbalch—Left the U.S. Coast Guard.

LT Michael T. Trippert—Transferred to U.S. Coast Guard Air Station Kodiak, Alaska.

CWO2 F. Scott Clendenin—Commanding Officer, U.S. Coast Guard Station Yaquina Bay, Oregon. Retired and lives in the Pacific Northwest.

CWO2 Robert Coster—Commanding Officer, Neah Bay, Washington, station. Transferred to U.S. Coast Guard Group Portland, Oregon.

CWO2 Thomas Doucette—Served at Group Portland, Oregon. Retired and living in the Pacific Northwest.

CWO2 Randy Lewis—Commanding Officer, U.S. Coast Guard Station Grays Harbor, Washington. Retired and lives in the Pacific Northwest.

BMCM George A. LaForge—Retired and lives in the Pacific Northwest.

BMCS Daniel E. Shipman—Transferred to Anchorage, Alaska. Promoted to BMCM and transferred to officer-in-charge, Tillamook Bay, Oregon. Retired and living in Alaska.

BMCS David Meyrick—Officer-in-charge, Quillayute River station.

BMC Tom Karczewski—Promoted to BMCS and transferred to U.S. Coast Guard Station Cape Disappointment, Washington, serving as executive petty officer.

BMC Steve Huffsteadler—Retired and living in the Great Lakes region.

BM1 Jonathan A. Placido—Transferred to a cutter in the Great Lakes.

BM1 Michael Saindon—Executive petty officer, Quillayute River Station. Promoted to BMC. Transferred to officer-in-charge, U.S. Coast Guard Station Morro Bay, California.

ASM1 Charles Carter—Serving at U.S. Coast Guard Air Station, Port Angeles, Washington.

QM1 Joseph N. Sekerak—Promoted to QMC and serving at Group Port Angeles.

BM2 Donald Meterko—Transferred to U.S. Coast Guard Air Station, Port Angeles, Washington. Transferred to Station Tillamook Bay, Oregon. Promoted to BM1. Made surfman.

BM2 W. Brent Cookingham—Transferred to U.S. Coast Guard station, Morro Bay, California. Retired and living in the Pacific Northwest.

BM2 Frank Heibert—Serving on the Aids to Navigation Team, Astoria, Oregon.

MK2 Thomas Byrd—Retired and living in the southeastern United States.

DC2 Michael Keller—Left the U.S. Coast Guard and living in Kansas.

AM3 Neal Amos—Left the U.S. Coast Guard and living in the southeastern United States.

BM3 Marcus M. Martin—Transferred to U.S. Coast Guard Station Grays Harbor, Washington. Promoted to BM2. Made surfman.

BM3 Paul R. Lassila—Left the U.S. Coast Guard and living in the Pacific Northwest.

MK3 James C. Johnston—Left the U.S. Coast Guard and living in the southeastern United States.

TC3 Gina Marshall—Serving at U.S. Coast Guard Group Port Angeles. Promoted to TC2.

Chris Koech—Left the U.S. Coast Guard and living in the Middle Atlantic states.

SN John A. Stoudenmire III—Left the U.S. Coast Guard and living in Alaska.

SN Jacques P. Faur—Transferred. Attended service school. Promoted to SK2 and serving at Grays Harbor station.

SA Benjamin F. Wingo—Transferred. Attended service school. Serving in Hawaii. Promoted to AVT3.

FA Zandra L. Ballard—Transferred. Attended service school. Serving in Guam. Promoted to TC3.

FA Falicia M. Brantley—Transferred. Attended service school. Promoted to FS3. Serving in Seattle, Washington. Married John DeMello.

FA John D. DeMello—Left the U.S. Coast Guard. Living in Seattle. Married Falicia Brantley.

Charles A. Lindenmuth II—Continues to serve on the Forks, Washington, Ambulance Corps.

Roseanne Lindenmuth—Continues to serve on the Forks, Washington, Ambulance Corps.

Appendix 2

Awards Received by the Major People in This Story

U.S. Coast Guard Station Quillayute River, Washington

Crew of Motor Lifeboat CG 44363

BM2 David A. Bosley—U.S. Coast Guard Medal (posthumously) (The highest award for valor in peacetime a person at a small-boat rescue station may receive.)
MK3 Matthew E. Schlimme—U.S. Coast Guard Medal (posthumously) (The highest award for valor in peacetime a person at a small-boat rescue station may receive.)
SN Clinton P. Miniken—U.S. Coast Guard Medal (posthumously) (The highest award for valor in peacetime a person at a small-boat rescue station may receive.)
SA Benjamin F. Wingo—U.S. Coast Guard Medal (The highest award for valor in peacetime a person at a small-boat rescue station may receive.)

Crew of Motor Lifeboat CG 44393

BM1 Jonathan A. Placido—U.S. Coast Guard Achievement Medal
MK2 Thomas L. Byrd—U.S. Coast Guard Achievement Medal
BM3 Marcus M. Martin—U.S. Coast Guard Achievement Medal
SN John A. Stoudenmire III—U.S. Coast Guard Achievement Medal

U.S. Coast Guard Air Station Port Angeles, Washington

Crew of Helicopter 6589

CDR Paul A. Langlois—Distinguished Flying Cross (The highest award a person in aviation can receive for valor in peacetime.)
CDR Raymond J. Miller—Distinguished Flying Cross (The highest award a person in aviation can receive for valor in peacetime.) This is the second Distinguished Flying Cross awarded to Commander Miller.
AM3 Neal Amos—Distinguished Flying Cross (The highest award a person in aviation can receive for valor in peacetime.)
ASM1 Charles Carter—U.S. Coast Guard Commendation Medal

U.S. Coast Guard Air Station Astoria, Oregon

Crew of Helicopter 6003

CDR Michael A. Neussl—U.S. Coast Guard Commendation Medal
LT Michael T. Trimpert—U.S. Coast Guard Achievement Medal
AM2 Richard J. Vanlandingham—U.S. Coast Guard Achievement Medal
ASM2 James Q. Lyon—U.S. Coast Guard Achievement Medal

Appendix 3

The Crew of U.S. Coast Guard Station, Quillayute River, Washington, 11–12 February 1997

BMCM George A. LaForge, Officer-in-Charge
BM1 Jonathan A. Placido, Executive Petty Officer
FA Zandra L. Ballard
FS1 Bruce E. Berkemeyer
BM2 David A. Bosley
SA Falicia M. Brantley
MK2 Thomas L. Byrd
BM2 W. Brent Cookingham
FA John D. DeMello
SN Jacques P. Faur
MK3 James C. Johnston
DC2 Michael W. Keller
FS2 Jason L. Koperski
BM3 Paul R. Lassila
BM3 Marcus M. Martin
FN William C. Matthews
SN Clinton P. Miniken
BM2 Donald J. Miterko
MK1 Bruce A. Mumford

MK3 Matthew E. Schlimme
SA Trevor K. Sowder
SN John A. Stoudenmire III
SA Benjamin F. Wingo
SN Sara L. Zurflueh

Notes

Chapter 1. The Fortress of Solitude

1. U.S. Coast Guard headquarters has decided that the new name for the small-boat stations will be "multimission station." The senior people at most of the coastal stations prefer to call them motor lifeboat stations, but I will use the term recognized by most Americans.

2. The other two are Neah Bay, Washington, Quillayute River's neighboring station to the north, and Grand Isle, Louisiana.

3. Richard L. Williams, *The Northwest Coast*, 23.

4. Office of Financial Management, *1999 Data Book*, 196–97, 220–21.

5. When enlisted personnel are in charge of a small-boat station, the person in charge is known as the officer-in-charge (OIC), the second in charge—the executive officer—has the title of executive petty officer (XPO), and the engineering petty officer is designated EPO. When commissioned officers are in charge, they hold the title of commanding officer (CO); if they are second in command they are the executive officer (XO); a commissioned engineering officer is an EO.

6. "Living in Balance: The Story of the Quileute Tribe," material gathered from the Internet at <http://www.arts.state.tx/housing/visual/htm>.

7. Tim McNulty, *Olympic National Park*, 197–99.

8. "Living in Balance."

9. McNulty, *Olympic National Park*, 197–99.

10. Telephone conversation, author with Barb Maynes, Olympic National Park public information officer, 6 July 1999.

11. Ranks and rates are those of February 1997 or of the date of interviews.

12. The author worked in a prison for two and a half years in the coastal area of Clallam County to the northeast of the Quillayute River station and can recall

having flashlights and candles constantly at the ready during the winter months at his residence outside the prison.

Chapter 2. Welcome Aboard

1. CWO2 Thomas D. Doucette was stationed at Quillayute River when the new station was put into commission. He related that "one of the reasons the station was moved was because there would be a television system that would monitor the bar. It was too hard to maintain, so it was discontinued." Information on the date of establishment and the moving of the station is from BMCM George A. LaForge, officer-in-charge of the Quillayute River station, June 1996.

2. CWO2 Doucette said that the artist, when finished, placed his name on the mural, then turned to Tom and said, "What do you think of it?" "Nice, Terry, but you spelled your name wrong," which brought forth an expletive.

Chapter 3. Boat Drivers

1. U.S. Coast Guard, *Boat Crew Qualification Guide: Coxswain*, II-6–3–II-6–8. Manual has updated changes entered in 1993.

2. U.S. Coast Guard, *44' MLB Operator's Handbook*, 4–1.

3. U.S. Coast Guard, *Boat Crew Qualification Guide: Surfman*, II-1–23–II-124.

Chapter 4. Port Duty Section

1. All information on Bosley's rescue comes from Scott Sunde, "Coast Guardsman remembered for heroism in rescue," *Seattle Post-Intelligencer*, 19 February 1997, 1, A5.

Chapter 5. Motor Lifeboats and Surf

1. On the island there are also some strong lights known as bar lights. The Quillayute River station is the only unit on the coast of Washington State to have bar lights. On 11 and 12 February 1997, however, the bar lights were not working. The lights were old, expensive to repair, and in a difficult position to maintain. After consulting with his staff and Master Chief LaForge about the lights, and receiving a quote on how expensive it would be to repair the lights, Group Commander Captain Philip C. Volk decided not to repair the lights, with the district staff not demurring on the decision. After the deaths, the lights were once again put into operation. Later, there was some questions as to whether the lights might have prevented the deaths. Given the low visibility and high seas, it is difficult to see how the lights might have prevented the deaths. Philip C. Volk, e-mail to author, 30 December 2000.

2. The U.S. Coast Guard's small-boat numbers on the bows of the craft begin with their length. Thus, CG 44393 is a 44-footer, boat number 393.

3. Background material on the history of lifeboats and the motor lifeboats of the U.S. Coast Guard are found in Dennis L. Noble, *Lifeboat Sailors: Disasters, Rescues, and the Perilous Future of the Coast Guard's Small Boat Stations*, 85–89.

The 44-foot motor lifeboat is able to right itself by means of fundamental structural design: compartmentation, layout, and light weight superstructure combined with a low center of gravity and centerline fuel-tank-free surface effect in capsize position. Self-bailing of the midships cockpit is achieved through 4–inch diameter non-return ball-check scupper valves. The self-righting capability of the new 47-foot motor lifeboat is achieved by a low center of gravity and the large volume of the enclosed watertight deckhouse. When the boat is in the inverted position the hull is resting on the deckhouse, which causes instability, which in turn results in self-righting. Information supplied by William D. Wilkinson, director emeritus of The Mariner's Museum and internationally recognized expert on the small-boat rescue craft of the U.S. Coast Guard, e-mail to author, 3 January 2001.

4. Tom was promoted to Chief Warrant Officer in 1997.

5. Material on the *Miss Renee* case comes from copies of the radio transmissions of the boats involved in the case, the official report on the case, and interviews by the author. For a detailed version of the case, see Dennis L. Noble, "A Sunday Evening in the Pacific Northwest," 4–10.

6. K. Adlard Coles, ed., *Heavy Weather Sailing*, 290.

7. Ibid., 298.

8. William G. Van Dorn, *Oceanography and Seamanship*, 187.

9. Laurence Draper, "Freak Waves," in Coles, *Heavy Weather Sailing*, 305.

10. Ibid., 306.

11. Ibid., 305.

Chapter 6. This Is a Drill

1. A boatswain's mate at the Ocean City, Maryland, station told me of a Mayday call they received. The SAR alarm went off. In his dash to the ready boat, the boatswain's mate turned his ankle. As soon as the watchstander could sort things out, he learned that the call came from a new boater who thought you prefaced any call to the U.S. Coast Guard with the word "Mayday."

2. *44' MLB Operator's Handbook*, 2–20.

3. Upon his retirement in 1999, Master Chief LaForge received from the crew of the Quillayute River station the glass from the windshield, framed and with an etching upon it depicting lifesaving.

Chapter 7. U.S. Coast Guard Group/Air Station, Port Angeles, Washington

1. *Peninsula Daily News*, 6 July 1998, A3.

2. Commander Langlois became commanding officer of Air Station Savannah,

Georgia, in 1997. During this tour of duty he was promoted to captain; in 1999 he was transferred to U.S. Coast Guard Headquarters and is serving as Chief, Enlisted Personnel Management Division. Material on Langlois's career is from his prepared biography (n.d, n.p.).

3. Commander Miller became executive officer of the Port Angeles Air Station and in 1999 became the commanding officer of Air Station Savannah, Georgia. Material on his career is from an e-mail from Miller to the author, 7 June 1999.

4. E-mail from Lieutenant David C. Billburg to author, 24 June 1999.

5. Arthur Pearcy, *U.S. Coast Guard Aircraft since 1916*, 107.

6. Quoted ibid., 112.

7. Ibid., 108.

8. "The HH-65A Dolphin," U.S. Coast Guard Fact Sheet (n.d., n.p.).

9. Ibid.

10. Capt. David Kunkel, U.S. Coast Guard Headquarters, Washington, D.C., e-mail to author, 1 July 1999.

11. Ibid.

12. Ibid.

13. In the pursuit of gender-neutral titles, the U.S. Coast Guard has since changed the name of this rate to aviation survival technician (AST).

14. ASTCM Keith R. Jensen, Helicopter Rescue Swimmer Program manager, U.S. Coast Guard Headquarters, Washington, D.C., e-mail to author, 8 July 1999.

15. See Barrett Thomas Beard, *Wonderful Flying Machines: A History of U.S. Coast Guard Helicopters*.

Chapter 8. Evening

1. LCDR Fred Myer, surface operations officer, Group Port Angeles, Washington, telephone conversation with author, 3 January 2001.

2. Times, weather, and locations of the *Gale Runner* prior to the call for help are unknown, as Schlag did not respond to letters concerning the case. The official investigation into the incident has very little testimony from Schlag. Why the board did not ask more questions is unknown.

3. Material on Schlimme attributed to his sister and early life is found in Leyla Kokmen, "Lost Guardsmen: kind, well qualified, 'loved driving boats,'" *Seattle Times,* 13 February 1997, news clip in the files of the Thirteenth Coast Guard District Public Affairs Office.

4. The *Mermaid* case is described in Robert Erwin Johnson, *Guardians of the Sea: History of the United States Coast Guard, 1915 to the Present*, 316–17. A detailed account in Darrell J. Murray's self-published scrapbook contains, among other items, the official investigation into the case. The scrapbook is unpaginated

and undated (1991?) and appears to be made up of photocopied material bound by Murray. Copy in author's files.

Chapter 9. "Mayday! Mayday!"

1. All references to the radio conversations in this account come from six audio tapes made from the tape recordings of the radio transmissions received during the case in the communications center of Group Port Angeles, Washington, on 11–12 February 1997; the tapes will be donated to the Quillayute River station.

2. Scott Sunde, "Sailboat Owner Tells of Tragedy on a Stormy Night," *Seattle Post-Intelligencer* (15 February 1997), news clipping found in scrapbook located in the files of the U.S. Coast Guard station, Quillayute River, Washington. It should be noted that the material in the newspaper did not make its way into the administrative investigation conducted by the U.S. Coast Guard.

3. As will be seen in part 3, there is a great deal of confusion in the official records. The material on what Master Chief LaForge said to Bosley is the first indication of this confusion. It will be shown that there were three investigations into the deaths: two—the safety investigation and the Commandant's Vessel Safety Board—are classified; the third—the administrative investigation—is open to the public. The administrative investigation notes that LaForge told Bosley to go to the bar and check out the conditions. I have received a letter dated 18 August 1997 in which VADM James M. Loy, the chief of staff of the U.S. Coast Guard, quotes material from the Commandant's Vessel Safety Board in which it is stated: "When the OINC [officer-in-charge, LaForge] made radio contact with the coxswain [Bosley], the OINC directed him to go to the bar and try to cross it. (To the OINC that meant to evaluate the conditions and if it was safe then proceed.) . . . Open ocean conditions outside the bar were much worse than what the OINC perceived. Based on past experience, the OINC felt that the coxswain, if he could safely cross the bar, could handle the perceived open ocean conditions." Part of the problem in interpreting what happened on 12 February is the U.S. Coast Guard's policy of not releasing material to researchers, but breaking their own rules in their correspondence and training aids. Material on LaForge's comments about crossing the bar is from Chief of Staff's Final Action Decision Letter of a Class "A" Mishap; Loss of Station Quillayute River MLB 44363 and Subsequent Death of Three Coast Guardsmen on 12 February 17, Serial 5520, 7 August 1997, with enclosure.

4. Christine Clarridge, "Only One Survived," *Seattle Times*, 30 May 1999, A-10.

5. Ibid.

6. Ibid.

7. Ibid.

8. Either the investigative board transcribed wrong, or Wingo misspoke, as there is no buoy number 3; there is, instead, daymarker number 3.

9. Clarridge, "Only One Survived," A-10.

10. Ibid.

11. Ibid.

Chapter 10. "I'm going!"

1. Martin's daughter, Kathyleen, was born 10 March 1997. He and Terise have since had another daughter, Kindra, born 15 August 1998.

Chapter 13. "We don't die, we save people"

1. Unless otherwise noted, all material on the work of emergency medical technicians Charles A. Lindenmuth and Roseanne Lindenmuth is in Charles A. Lindenmuth II, "Rescue the Rescuer," an unpaged, undated typescript in the files of the U.S. Coast Guard station, Quillayute River, Washington.

Chapter 15. "Everyone's heart just sagged"

1. Zandra Ballard remembers that on the weekend before the deaths, "the Master Chief, me, and three others of the crew went to a memorial service for one of the local fishermen who had drowned. Somehow taps was brought up between us. I can remember Master Chief telling us that he had never lost any of his crew in the line of duty. He had crew that had died, but never in the line of duty."

2. Master Chief Petty Officer of the Coast Guard Vincent W. Patton III, e-mail to author, 27 January 2000. Once again, when I could not get a response from headquarters, he took time to explain CISM to me.

3. All quotes from Wingo, unless otherwise noted, are in Clarridge, "Only One Survived."

4. Janice J. Maxson, "Death and the Press—A Case Study: Media Coverage of the Coast Guard Tragedy at Station Quillayute River, La Push, Washington, February 12th 1997," a typescript prepared for the University of Washington, School of Communications, 1 April 1997, located in the files of the U.S. Coast Guard station Quillayute River, Washington, 1.

5. Ibid.

6. Ibid., 8.

7. Ibid., 9.

8. Ibid.

9. There was no media interview of Wingo until 30 May 1999, in the *Seattle Times.*

10. Maxson, "Death and the Press," 10.

11. Ibid., 8.

12. The CG 44363 was cut into three pieces and removed from the cove on James Island. The section containing the aft survivor's compartment now sits atop a large

pile of material in a recycling yard off Highway 101, about halfway between Sequim and Port Angeles, Washington. The author views it every time he drives into Port Angeles.

Chapter 16. Investigations

1. Photocopy of a two-page document titled "MLB 44363 Rollover Mishap Analysis Board (MAB) Feedback," undated and unpaginated, sent to the author from an unknown source.

2. Unless otherwise noted, all material from Senior Chief Shipman's Page 7 remarks are contained within the administrative investigation.

3. Other senior enlisted people in the U.S. Coast Guard have pointed out how hard it is to relieve someone for cause because too many restrictions are put upon the officer-in-charge.

4. Long quotation is from the Historian of the U.S. Coast Guard's Internet web site: <http://www.uscg.mil/hq/g-cp/history/collect.html>.

Chapter 17. Questions

1. Letter, "Chief of Staff's [VADM James M. Loy] Final Decision Letter of a Class 'A' Mishap; Loss of Station Quillayute River MLB 44363 and Subsequent Death of Three Coast Guardsmen on 12 February 1997," serial 5102, 21 April 1997, 2–6.

2. See note 1, chapter 5.

3. Letter, RADM Ernest R. Riutta, assistant commandant for operations, U.S. Coast Guard Headquarters, Washington, D.C., serial 5730, 14 February 2000, with enclosures, to author, enclosure, 2.

4. *The Herald* (Snohomish and Island Counties, Washington), 19 June 1997, located in the files of the Public Affairs Office, Thirteenth Coast Guard District, Seattle, Washington.

5. (Seattle) *Post-Intelligencer,* 20 June 1997, C-1, C-4. Note that by two days after the release of the investigation, the story had been put back to the "C" section of the Seattle newspaper.

6. E-mails between CDR John Philbin and author, 23 November 1999, 9 and 10 January 2000.

7. Ibid.; e-mails between Master Chief Petty Officer Vincent W. Patton III and author, 7 and 14 February 2000.

8. There are many similarities between the Quillayute River deaths and the crashing of the Humboldt Bay helicopter. In each case, a sailboat was in distress during a severe storm at night. At Quillayute River, a helicopter performed the rescue of the people from the sailboat. In the Humboldt Bay case, a surface unit, the cutter *Edisto,* a 110-foot patrol boat, performed the rescue of the people aboard the

sailboat, while the people aboard the helicopter perished. The sailboat in the Humboldt Bay case suffered a knockdown, and the crew took to a life raft. The investigation brought out the inexperience of the helicopter crew. The Humbolt Bay helicopter had the following experience levels: The copilot was a recently designated copilot and had flown over 43 hours in the past nine weeks since he had started flying at Humboldt Bay. During that period, he had "only one instrument approach (with no actual or simulated instrument time logged), less than an hour (0.8) of nighttime (shortly after sunset) and night over-water instruments." This was the first SAR case for the pilot in command as an aircraft commander; he had held the designation for six months. The flight was the copilot's first night SAR case, and the rescue swimmer's first SAR case. Information on the Humboldt Bay case, which consisted of the message on the final results of the case, was sent to me from an unidentified source.

9. Riutta to Noble, enclosure, 2.

10. Ibid., 1.

11. Ibid.

12. *Seattle Times*, 30 May 1999, A-11.

13. Riutta to Noble, enclosure, 1.

14. Routine U.S. Coast Guard message sent on 2 June 2000. Subject: Boat crew safety belt hooks. Copy of message sent to me from an unknown source.

15. The comments were from the staff of the former commandant, Admiral Robert E. Kramek. The date of the first 47-footer at Cape Disappointment came from an e-mail from Lieutenant Daniel C. Johnson, commanding officer of Cape Disappointment station, to author, 6 June 2000.

16. "Public address given by Rear Admiral Blayney, Remarks at USCGC OSPREY Commissioning, Port Townsend, Washington," n.p. Speech is found on the Thirteenth Coast Guard District web page, <http://www.uscg.mil/d13/>.

17. E-mail, BM1 Michael Saindon to author, 13 January 2000.

18. Riutta to Noble, enclosure, 1.

19. Ibid., 2.

20. Ibid.

21. E-Mail, Philip C. Volk to author, 16 November 1999.

22. Admiral James Loy, "State of the Coast Guard," *Coast Guard* (June 1999): 18.

23. All material on the *Morning Dew* case is found in William Dean Lee, "SAR Case Study: S/V Morning Dew," *On Scene: The Journal of U.S. Coast Guard Search and Rescue* (Summer 1999): 14–15.

24. Riutta to author, enclosure, 4.

25. Ibid.

26. Loy, "State of the Coast Guard," 19.
27. Riutta to Noble, enclosure, 3.
28. Ibid., 5.

Chapter 18. Causes

1. See note 3, chapter 9, for additional material on what one of the classified investigations has to say about what LaForge passed to Bosley. The information cited in the narrative for this chapter is from the administrative investigation.
2. Dana A. Goward, "Life-Saving Service Left in the Cold," 52.
3. Riutta to Noble, enclosure, 4.
4. ALCOAST [All Coast Guard] 097/95 COMDTNOTE 7100, 1646Z OCT 95, sections 1 and 2, located in the files of the Historian of the U.S. Coast Guard, U.S. Coast Guard Headquarters, Washington, D.C.
5. Kirk Moore, in the *Asbury Park Press* (New Jersey), 25 February 2000, copy sent to author from unidentified source.
6. Photocopy of article sent to author from an unidentified source.
7. Material on personnel strength at Quillayute River comes from the Unit Personnel Data Report, various dates, located in the files of the Quillayute River station.
8. See Johnson *Guardians of the Sea,* 318.
9. Riutta to Noble, enclosure, 2.
10. This comment is in "First Endorsement on CDR J. M. Hasselbalch ltr. 5830 of 19Mar97," signed by Rear Admiral J. David Spade, dated 28 March 1997, 1–2.
11. Riutta to Noble, enclosure, 3.
12. Ibid., 3.
13. Robert F. Bennett, "The Life-Savers, 'For Those in Peril on the Sea,'" 63.
14. Reminiscences of Admiral Willard J. Smith (Annapolis, Md.: U.S. Naval Institute, 1 April 1971),71.
15. Dennis L. Noble, *That Others Might Live: The U.S. Life-Saving Service, 1878–1915* (Annapolis, Md.: Naval Institute Press, 1994), 63–70.
16. Ironically, AM3 Neal Amos, the helicopter crewman who won the Distinguished Flying Cross in the *Gale Runner* rescue, was forced out of the U.S. Coast Guard by this up-or-out policy.
17. For feelings of people at stations, see, Noble, *Lifeboat Sailors,* passim. and especially 184–207. For the feelings of World War I troops in the trenches, see Anthony Kellett, *Combat Motivation: The Behavior* of *Soldiers in Combat* (Boston: Kluwer-Nijhoff, 1982), 119–22.
18. All material on the budget comes from *U.S. Coast Guard: Fiscal Year 2000 Budget in Brief* (Washington, D.C.: Commandant, U.S. Coast Guard, n.d.).

19. Riutta to Noble, enclosure, 5.

20. This quotation was found on 20 February 2000 on the Internet at <www.uscg.mil/hq/g-a/healy/intro.htm>.

21. E-mail to Dennis L. Noble, 16 February 2000. A proposed "History of the Marine Safety Office in Tampa, Florida."

22. Christopher Forando, "Special Ops Needs a New Player," U.S. Naval Institute *Proceedings* 123/10/1,136 (October 1997): 44.

23. This passage was located in a briefing outline for the Quillayute River deaths. The briefing outline is in the "Quillayute River" file located in the office of the Historian of the U.S. Coast Guard, U.S. Coast Guard Headquarters, Washington, D.C.

24. Riutta to Noble, enclosure, 5.

Chapter 19. Lessons Learned

1. An unsigned, undated letter, serial 5730, to Dennis L. Noble, in an e-mail enclosure from Commander John Philbin, press assistant to the commandant, on 14 February 2000, in which Philbin stated: "Attached is Admiral Loy's response. A signed copy is in the mail." The dated and signed copy reached the author on 23 February 2000.

2. The oldest scenario was sent to me by an unknown source, the newest from the Chief Petty Officer Academy located at the U.S. Coast Guard Academy, New London, Connecticut.

3. Adam Katz-Stone, "What really happened the night disaster hit?" *Navy Times* (August 25, 1997), 20. Unless otherwise noted, all quotes hereafter in this section are from this article.

Chapter 20. The Light at the End of the Tunnel?

1. "PROJECT KIMBALL: Executive Steering Committee, Guidance Team, Project Team, Charter." An undated typescript sent to Dennis L. Noble from an unknown source, 1. All material on Project Kimball comes from this document, unless otherwise noted. The project is named after Sumner I. Kimball, the man most responsible for the establishment and smooth running of the old U.S. Life-Saving Service.

2. Goward, "Life-Saving Service Left in the Cold," 52.

3. Ibid., 56.

4. Ibid.

5. Ibid.

6. Letter, William D. Wilkinson to Dennis L. Noble, 14 February 2000.

7. Goward, "Life-Saving Service Left in the Cold," 56.

8. E-mail, Dana A. Goward to Dennis L. Noble, 25 April 2000, 1.

9. Ibid.

10. Ibid.

11. Milligan, "Comments and Discussion," U.S. Naval Institute *Proceedings* 126/2/1,164 (February 2000): 18, 20.

12. Ibid., 20.

13. Alex Larzelere, *Coast Guard at War: Vietnam, 1965–1975* (Annapolis, Md.: Naval Institute Press, 1997), 7.

14. Ibid., 15.

Chapter 21. Reflections

1. To be objective, there continues to be a debate within the U.S. Coast Guard as to just what it is supposed to be. This debate is not just among the officer corps and not just in the halls of headquarters; it is service-wide and at all levels among officers and enlisted men. One has only to read the comments posted throughout the year on Fred's Place, an Internet site devoted to matters concerning the U.S. Coast Guard, to sample this debate. To sample this debate, see www.fredsplace.org.

2. In a previous book on small-boat stations, I sent letters to eight district offices with a request to have an answer back within three months. Here is the scorecard: Two districts responded promptly with information; one said, "we want to work with you" and that was the last heard from them. One district said three months was too quick to get a reply back and one district replied a month late, while another said they lost the request and replied that they would respond to another request, which they apparently lost again, as that was the last time they corresponded. Finally, two districts did not bother to respond at all. This dismal record carried on upward to Admiral Robert Kramek, then commandant of the U.S. Coast Guard, who did not bother to respond to a letter requesting an interview.

3. Clarridge, "Only One Survived," A-1.

4. Ibid., A-11.

5. Ibid., A-1.

6. Ibid., A-11.

7. Maxson, "Death and the Press," 11.

8. Lindenmuth, "Rescue the Rescuer," 5.

9. BMCS Murray's attempts to obtain the transcript of the investigation into the *Mermaid* case is given in a self-published scrapbook, *First Reunion on 30th Anniversary Triumph-F/V Mermaid Incident.*

Postscript

1. U.S. Coast Guard News, Ninth Coast Guard District, 26 March 2001; "Taps," *U.S. Coast Guard Reservist* (May 2001): 26.

2. Port Angeles (Washington) *Peninsula Daily News,* 6 March 2001, A-1, A-4.

3. U.S. Coast Guard news releases, 22 March 2001 and 3 April 2001.

4. *Peninsula Daily News,* 7 March 2001, A-1.

5. Ibid., A-2.

6. Ibid., 9–10 March 2001, A-1, A-5.

7. E-mail to Dennis L. Noble from CWO Chris Haley, public affairs officer, Coast Guard District Thirteen, 20 April 2001.

8. Ibid.

9. Ibid.

Selected Bibliography

Files of the Historian of the U.S. Coast Guard, U.S. Coast Guard Headquarters, Washington, D.C.

Files of the U.S. Coast Guard Station, Quillayute River, LaPush, Washington

"Investigation into the Capsizing and Subsequent Loss of MLB 44363 and the Death of Three Coast Guard Members That Occurred at Coast Guard Station Quillayute River on 12 Feb 1997," serial 5830, 19 March 1997, with endorsement 28 March 1997 by the Commander, Thirteenth Coast Guard District.

"Action of the Final Reviewing Authority on the Investigation into the Capsizing and Subsequent Loss of MLB 44363 and the Death of Three Coast Guard Members That Occurred at Coast Guard Station Quillayute River on 12 February 1997," signed by the Commandant of the U.S. Coast Guard on 14 June 1997.

U.S. Government Publications

Coast Guard. Washington, D.C.: U.S. Coast Guard, various dates.

On Scene: The Journal of U.S. Coast Guard Search and Rescue. Washington, D.C.: U.S. Coast Guard, various dates.

U.S. Coast Guard. *Boat Crew Qualification Guide: Coxswain.* Vol. 2. Washington, D.C.: U.S. Coast Guard, 1986.

———. *Boat Crew Qualification Guide: Surfman.* Washington, D.C. [?]: U.S. Coast Guard, 1989.

———. *44' MLB Operator's Handbook.* Washington, D.C. [?]: U.S. Coast Guard, 1992.

Oral History

Reminiscences of Admiral Willard J. Smith, U.S. Coast Guard (Retired). Annapolis, Md.: U.S. Naval Institute, 1 April 1971.

Newspapers

The Herald (Snohomish and Island Counties, Washington)
Navy Times
Peninsula Daily News (Port Angeles, Washington)
Post-Intelligencer (Seattle)
Seattle Times

Secondary Sources

Beard, Barrett Thomas. *Wonderful Flying Machines: A History of U.S. Coast Guard Helicopters.* Annapolis, Md.: Naval Institute Press, 1996.

Bennett, Robert F. "The Life-Savers: 'For Those in Peril on the Sea.'" U.S. Naval Institute *Proceedings* 102/3/877 (March 1976): 54–63.

Coles, K. Adlard. *Heavy Weather Sailing,* 3d ed. Clinton Corners, N.Y.: John de Graff, 1981.

Forando, Christopher. "Special Ops Needs a New Player." U.S. Naval Institute *Proceedings* 123/10/1,136 (October 1997): 44–45.

Goward, Dana A. "Life-Saving Left Out in the Cold." U.S. Naval Institute *Proceedings* 125/12/1,162 (December 1999): 52–56.

Information Please Almanac. New York, Dan Goldenpaul Associates, 1969.

Johnson, Robert Erwin. *Guardians of the Sea: History of the United States Coast Guard, 1915 to the Present.* Annapolis, Md.: Naval Institute Press, 1987.

Larzelere, Alex. *The Coast Guard at War: Vietnam, 1965–1975.* Annapolis, Md.: Naval Institute Press, 1997.

McNulty, Tim. *Olympic National Park.* Boston: Houghton Mifflin Company, 1996.

Milligan, Michael P. "Comment and Discussion." U.S. Naval Institute *Proceedings* 126/2/1,164 (February 2000): 18, 20.

Murray, Darrell. *First Reunion on 30th Anniversary Triumph—F/V Mermaid Incident, 12 January 1961–12 January 1991, Astoria, Oregon.* N.p., 1961.

Noble, Dennis L. *That Others Might Live: The U.S. Life-Saving Service, 1878–1915.* Annapolis, Md.: Naval Institute Press, 1994.

———. *Lifeboat Sailors: Disasters, Rescues, and the Perilous Future of the Coast Guard's Small Boat Stations.* Washington, D.C.: Brassey's, 2000.

———. "A Sunday Evening in the Pacific Northwest." *Wreck & Rescue: The Journal of the U.S. Life-Saving Service Heritage Association* 13 (Winter 2000): 4–10.

Office of Financial Management. *1999 Data Book.* Olympia, Wash.: Office of Financial Management, 1999.
Pearcy, Arthur. *U.S. Coast Guard Aircraft since 1916.* Annapolis, Md.: Naval Institute Press, 1991.
Van Dorn, William G. *Oceanography and Seamanship.* New York: Dodd, Mead, & Company, 1974.
Williams, Richard L. *The Northwest Coast.* Alexandria, Va.: Time-Life, 1973.
World Almanac, 2000. Mahwah, N.J.: World Almanac Books, 1999.

Index

Dennis L. Noble entered the U.S. Coast Guard in 1957 and retired in 1978 as a Senior Chief Petty Officer (E-8). After retirement, he returned to school and earned a Ph.D. in U.S. history at Purdue University. He is author of nine books, seven of them on the history of the U.S. Coast Guard.